Sustainability

Second Edition

Jkar

Key Concepts Series

Barbara Adam, *Time*
Alan Aldridge, *Consumption*
Alan Aldridge, *The Market*
Jakob Arnoldi, *Risk*
Will Atkinson, *Class*
Colin Barnes and Geof Mercer, *Disability*
Darin Barney, *The Network Society*
Mildred Blaxter, *Health 2nd edition*
Harriet Bradley, *Gender 2nd edition*
Harry Brighouse, *Justice*
Mónica Brito Vieira and David Runciman, *Representation*
Steve Bruce, *Fundamentalism 2nd edition*
Joan Busfield, *Mental Illness*
Margaret Canovan, *The People*
Andrew Jason Cohen, *Toleration*
Alejandro Colás, *Empire*
Patricia Hill Collins and Sirma Bilge, *Intersectionality*
Mary Daly, *Welfare*
Anthony Elliott, *Concepts of the Self 3rd edition*
Steve Fenton, *Ethnicity 2nd edition*
Katrin Flikschuh, *Freedom*
Michael Freeman, *Human Rights 2nd edition*
Russell Hardin, *Trust*
Geoffrey Ingham, *Capitalism*
Fred Inglis, *Culture*
Robert H. Jackson, *Sovereignty*
Jennifer Jackson Preece, *Minority Rights*
Gill Jones, *Youth*
Paul Kelly, *Liberalism*
Anne Mette Kjær, *Governance*
Ruth Lister, *Poverty*
Jon Mandle, *Global Justice*
Cillian McBride, *Recognition*
Anthony Payne and Nicola Phillips, *Development*
Judith Phillips, *Care*
Chris Phillipson, *Ageing*
Robert Reiner, *Crime*
Michael Saward, *Democracy*
John Scott, *Power*
Timothy J. Sinclair, *Global Governance*
Anthony D. Smith, *Nationalism 2nd edition*
Deborah Stevenson, *The City*
Steven Peter Vallas, *Work*
Stuart White, *Equality*
Michael Wyness, *Childhood*

Sustainability

Second Edition

Leslie Paul Thiele

polity

First edition published in 2013 by Polity Press
This edition published in 2016 by Polity Press

Polity Press
65 Bridge Street
Cambridge CB2 1UR, UK

Polity Press
350 Main Street
Malden, MA 02148, USA

ISBN-13: 978-1-5095-1106-8
ISBN-13: 978-1-5095-1107-5(pb)

A catalogue record for this book is available from the British Library.

Library of Congress Cataloging-in-Publication Data

Names: Thiele, Leslie Paul, author.
Title: Sustainability / Leslie Paul Thiele.
Description: Cambridge, UK ; Malden, MA : Polity Press, [2016] |
   Includes bibliographical references and index.
Identifiers: LCCN 2016009865 (print) | LCCN 2016015961 (ebook) |
   ISBN 9781509511068 (hardback) | ISBN 9781509511075 (pbk.) |
   ISBN 9781509511099 (Mobi) | ISBN 9781509511105 (Epub)
Subjects: LCSH: Sustainable development. | Sustainable development–
   Social aspects.
Classification: LCC HC79.E5 T4759 2016 (print) | LCC HC79.E5
   (ebook) | DDC 338.9/27–dc23
LC record available at https://lccn.loc.gov/2016009865

Typeset in 10.5 on 12 pt Sabon
by Toppan Best-set Premedia Limited
Printed in Great Britain by CPI Group (UK) Ltd, Croydon

For further information on Polity, visit our website:
politybooks.com

Contents

Preface

The first edition of this book was completed on July 1, 2012 in Nelson, British Columbia. July 1 is Canada Day, and the festivities were about to start: music, singing, dancing, games, speeches, food, and, at nightfall, fireworks. It is always a fine celebration that brings this small mountain community together. That year was no different, except...

Except the grounds of the city park had turned into a mud bog. It has been raining most of the day, and the just completed month of June was the wettest ever, doubling the previous record. The month of March had also set a record for precipitation.

Predictably, fewer people participated in the sodden event. And the fireworks had to be cancelled. The playing fields where people gather to watch the dazzling pyrotechnics when the music and dancing ends were flooded, merging with the waters of Kootenay Lake, which had breached its shore a few days back.

At the time, I hoped these record-setting weather events were just that: weather. More likely, I knew, we were seeing climate change at work. Now, as I write the preface to this second edition, scientists have just announced that the preceding year was the hottest the planet has known since records starting being kept in the 1800s. The urgency of our situation is palpable.

The world is quickly changing, and fortunately much of this change is for the good. Across the globe people are becoming informed, creatively solving problems, and collectively taking action. The second edition of *Sustainability* has been fully revised and updated to account for these trends. It integrates an expanded global breadth with increased attention to the importance of local relationships, responsibilities, and opportunities. New Inquire and Explore sections are provided at the end of each chapter to help readers probe and deepen central debates and topics. The thought- and discussion-provoking questions and issues presented here are directly linked to graphics, data, readings, podcasts, videos, and resource-rich websites at conservationandcreativity.net

There is still reason to worry that future generations living through the calamity of a much-altered climate – a calamity including extreme weather, rising and increasingly acidic oceans, countless species extinctions, food scarcity, coastal flooding, and the displacement of millions of people – will ask why their ancestors did so little to avert it. But I take heart that individuals and communities on our beautiful planet are learning how to live sustainably, and that there will be much to celebrate together in the future.

It is rightly said that sustainability is not a spectator sport. My hope is that the discussions, relationships, and actions fostered by this book, and the Inquire and Explore resources, will contribute to making that statement substantially truer.

Acknowledgments

This book was written with students in mind, specifically those enrolled in the Bachelor of Arts in Sustainability Studies and the Minor in Sustainability Studies at the University of Florida. These students are refreshingly diverse in their backgrounds, interests, and perspectives. I am grateful for their diversity as well as their abundant curiosity, energy, and dedication. I appreciate the astute editing of the manuscript provided by Mauro Caraccioli, Jonathan Davis, Maria Martinez, and Robyn Ostroff. Special thanks are due to Seaton Tarrant and Hugo Chaves for their additional work on the second edition.

Introduction

While the motivation to live sustainably dates back to ancient times, its vocabulary is but three decades old. In this relatively short span of time, the concept and practice of sustainability has generated lifestyle changes for individuals, innovations within business, design, engineering, and agriculture, historic policies and laws at municipal and state levels, and crucial international protocols and agreements. Sustainability is now advocated by a growing number of citizens, corporations, professional agencies, educational institutions, civic organizations, political parties, and governments. It is one of a very few ideals – joining the ranks of *democracy* and *human rights* – that receive near universal endorsement. Across a diverse globe of peoples and nations, sustainability increasingly provides a common language, a lingua franca for the twenty-first century. Indeed, sustainability has been dubbed a "megatrend," that is, a phenomenon with a massive and enduring impact on culture, the economy, politics, society, and technology.[1]

But what is sustainability? Is it an ethical ideal? Is it an ascendant ideology? Or is it a scientifically based effort to live well in an ever more crowded and complex world of increasingly scarce resources? Sustainability certainly has ethical components. It is grounded in moral claims about the responsibilities and obligations of individuals and organizations. And its ethics find support in age-old virtues.

Though seldom labeled an ideology, sustainability seems to fit the description. It constitutes a coherent set of interrelated beliefs and values that establish how collective life might better be organized. Some would consider sustainability the predominant ideology of our times, as it is global in its reach and ecumenical in its appeal.

To be sure, sustainability makes good use of science to ensure the adaptive management of scarce resources in a crowded and complex world. Scholars demonstrate that unsustainable lifestyles, economies, technologies, and social practices are hurtling us toward a precipice. In an age of global climate change, resource depletion, and "failing states" that can no longer meet their people's basic needs, sustainability has been given the daunting tasks of "rescuing civilization" and "saving the planet." Learning to live and work sustainably is arguably the greatest challenge of our times, and it demands the best science.

An adaptive art

For all its practical importance, the meaning of sustainability remains unsettled. Owing to the frequency and looseness of its usage, *sustainability* has been called "one of the least meaningful and most overused words in the English language."[2] The best response to this sort of criticism is not to stop using the word, but to define it clearly while making its practice more measurable and impactful.

Sustainability is most easily defined by saying what it is *not*. A practice, relationship, or institution is *not* sustainable if it undermines the social, economic, or environmental conditions of its own viability. It is unsustainable to extract water from rivers, lakes, and aquifers at a faster rate than they can be naturally recharged by rain and snow. Doing so will produce water-starved communities. Eroding the land upon which crops grow faster than fertile soil is naturally regenerated is not sustainable agriculture. It will end with failing farms and hunger. Running a corporation consistently in the red, with revenues that do not exceed expenses, is not sustainable business. It will end in bankruptcy. To be sustainable is to avoid collapse.

With this in mind, practitioners often define sustainability as meeting current needs in a way that does not undermine future welfare. This is a passable definition. But it neglects the crucial question of *whose* needs are being met and *whose* future welfare is at stake. As discussed in these pages, sustainability refers not only to the long-term survival of a specific practice, relationship, or institution. It entails an expanded scope. Sustainability extends our concern beyond the welfare of those participants who are directly involved in a practice, relationship, or institution. It also concerns the welfare of other stakeholders who become impacted by our actions – the welfare of people and other species distant in time or space.

In an increasingly interdependent world, virtually everyone is impacted – sooner or later – by everything we do. The consequences of our actions (and inactions) cross borders and generations, spanning the globe and casting long shadows into the future. The natural resources utilized in the goods we purchase in shops were likely extracted, processed, and assembled in dozens of other countries. The waste that we generate when these manufactured goods get thrown away may end up back in these far-off lands, or impact the health of our children's children. The carbon dioxide that today exits factory smokestacks, energy plants, and our vehicles' tailpipes will have its greatest impact on distant progeny, as the planet's temperature steadily rises under a blanket of greenhouse gases and oceans swamp shorelines. Sustainability concerns the global, long-term impact of our practices, relationships, and institutions because we live in a connected world. To live sustainably is to act with an expansive temporal and geographic awareness.

Sustainability is typically understood as the effort to use natural resources less wastefully. For many, it simply entails recycling and being energy efficient. But sustainability is both more challenging and more rewarding. It pushes us to better understand our world and ourselves, as we participate in expanding and shifting global networks. And it cultivates a sense of responsibility for maintaining the integrity and improving the health and resilience of the ecological, social, and economic networks that support us.

Notwithstanding common misperceptions, sustainability is not simply about preserving things. Sustainability requires change. It pertains equally to conservation *and* creativity. To be effective in our conservation efforts, we have to adapt to a changing world. But the change we foster must be limited in scope, and it must occur within an appropriate time frame. Practicing sustainability entails managing well the scale and speed of change.

For example, global warming and global cooling have occurred many times in the history of the planet. Planetary climate change is nothing new. However, the current level of atmospheric carbon dioxide, measured in parts per million, is growing at many times the rate that occurred the last time the earth lost its polar ice caps millions of years ago, when much of its landmass became submerged under enlarging seas. Our species is altering the climate of the planet at such an accelerated rate that it will prove impossible for millions of species, quite possibly including our own, to adapt in time. Along with habitat loss and pollution, global warming is already one of the top causes of species extinction, and scientists assert that we are only at the cusp of anthropogenic (human-caused) climate change. Our use of fossil fuels and our destruction of forests are unsustainable practices not because they are changing the climate, but because they are causing such disruptive change at a scale and speed that preempt successful adaptation by many forms of life.

Sustainability is not a recipe for creating a planetary museum where nothing ever changes. Its goal is not to achieve a "culture of permanence."[3] Rather, to practice sustainability is to manage change such that civilization does not undermine the conditions that allow it to flourish within a supportive web of life. Sustainability entails preserving crucial capacities and resources by way of adaptive change.

If we understand ideologies to be stubborn sets of beliefs that refuse revision in the face of good evidence, then sustainability is not an ideology. Optimally, it is an art that skillfully grounds moral commitments in, and adapts its practices to, the best available science. Sustainability is an adaptive art wedded to science in service to ethical vision. To define it succinctly: Sustainability is the practice of satisfying current needs without sacrificing future wellbeing by preserving core values and relationships while managing the scale and speed of change.

To maintain crucial relationships and values while embracing adaptive change requires the enduring and expansive integration of ecological health, economic welfare, social empowerment, and cultural creativity. Ecological health entails the conservation and cultivation of vibrant, biodiverse wild spaces, ecosystems, and pollution-free land, water, and atmospheric environments. Economic welfare entails the creation of infrastructure and opportunities for individuals and families to pursue enterprise and material prosperity, with mechanisms to avoid corrosive disparities of wealth and aid for those who cannot meet their basic needs. Social empowerment entails the creation of institutions and opportunities for people to meaningfully direct their individual and collective lives, including access to education, democratic government, basic human rights, and a vigorous civil society. Cultural creativity entails opportunities to engage, explore, enrich, and innovate in all facets of human culture, including science and knowledge, ethics and politics, economy and technology, customs and diet, arts and recreation, religion, and spirituality. Often, when people use the word *sustainability* they are thinking about environmental issues. To be truly sustainable, however, a practice, relationship, or institution must do more than protect nature and conserve natural resources. It must meet economic needs and cultivate economic opportunities. In turn, it must meet social needs and cultivate equitable relationships. You are not running a sustainable business, no matter how "green" your practices, if you consistently fail to make a profit and cannot pay your employees. Likewise, in today's connected world, social institutions that do not empower stakeholders and treat them equitably cannot contribute to enduring economic welfare and ecological health.

Sustainability has traditionally been described as standing on the three pillars of *society, ecology,* and *economy,* or, alternatively, as grounded on the "triple bottom line" of *people, planet,* and *profit.* The point is that these three goods stand – or fall – together. You cannot have one without the other two, owing to the interdependent nature of our world. But there is a problem with these formulas. They suggest that sustainability requires a static, rigid balance. By insisting on cultural creativity as a fourth component of sustainability, we underline the fact that our practices, relationships, and institutions have to initiate and respond to change if they are to endure for long. Sustainability demands imagination and innovation.

The humorist Will Rogers once remarked that even if you are on the right track, you will eventually get run over by a train if you just sit there. To practice sustainability we cannot just sit there. We have to learn and we have to adapt. Without creativity, it is impossible to sustain what we value.

A contested concept and practice

Sustainability is endorsed by a diverse and ever-growing number of citizens and governments, local, national, and international organizations, producers and consumers, clients and corporations. Sustainability is very popular today, even trendy. Popularity is a double-edged sword. To be sure, it is exciting and hopeful for growing numbers of people to embrace sustainability. But is it a good thing for sustainability to be championed by people who may have little understanding of its meaning, little intention of putting it into practice, and track records indicating that their lifestyles, policies, or business operations are anything but sustainable?

Virtually all politicians and rulers today, including the most autocratic, endorse democracy. Likewise, governments and businesses across the world endorse sustainability. But verbal endorsements often do not translate into policy or practice. Lip service does not move arms and legs into action. It is often said that hypocrisy is the compliment that vice pays to virtue. Perhaps, then, the widespread and growing endorsement of sustainability, even when preaching does not lead to practice, is still a victory of sorts. Perhaps it represents progress, an advance over a time when few even paid lip service to sustainability, and endorsements were rare.

Or perhaps not. Vague, hypocritical, or unsupported endorsements of sustainability may fatally weaken the concept and undermine its practice. Sustainability today often gets reduced to a slogan bandied about that results in few if any meaningful efforts or achievements. Many large corporations, for example, make claim to being "green" and embrace sustainability as a core value. But their business practices fall far short of anything they espouse. Indeed, the sustainability of their products or practices is largely a fabricated claim. Such "greenwashing" is a response to the perceived concerns of customers, clients, and stakeholders. As a megatrend, sustainability now serves as a marketing tool. To paint oneself green is savvy public relations; good PR, and often little more. The endorsement of sustainability serves as a green veneer for business as usual rather than a driver of fundamental change.

Governments also greenwash. Representatives of non-governmental organizations attending the 2012 Earth Summit in Rio de Janeiro, for example, declared the meeting an "epic failure" that has "given us a new definition of hypocrisy." The summit, which brought together more than a hundred heads of states and delegates from 188 nations, was mandated to deliver "a pathway for a sustainable century." But its officially endorsed document, entitled "The Future We Want," was widely criticized for failing to address major challenges or commit countries to specific actions and firm deadlines. "We didn't get the Future We Want in Rio," the activist group Greenpeace concluded, "because we do not have the leaders we need. The leaders of the most powerful countries supported business as usual, shamefully putting private profit before people and the planet."[4] The summit was described as another example in a 20-year trend of gaining widespread endorsement of sustainability "by sacrificing real substance."[5]

Since ancient Roman times, consumers in the marketplace have been given a warning: *caveat emptor!* It means *buyer beware*. Not everything that glitters is gold. People interested in sustainability might also heed this advice. Not everything that shimmers green is truly sustainable. One must scratch beneath the surface.

The word "sustainability" derives from the Latin *sustinere*, which literally means "to hold up." Something is sustainable if it endures, persists, or holds up over time. But we do not want all things to hold up over time. We want many destructive things to end, and to end sooner rather than later. The sad truth is that certain people and institutions are the cause of much harm. There is money to be made and power to be gained from economic instability, social injustice, ecological degradation, and ignorance – at least in the short term. For this reason, sustainability will never be endorsed by everyone. Some stand to gain from unsustainable practices, relationships, and institutions. We should not shy away from the fact that there are tough battles to be fought with individuals and organizations whose narrow self-interest leads them to oppose truly sustainable practices.

When sustainability gets defined in terms of general values and abstract concepts without specific means of implementation or a sense of the stakes in the game and the opponents to be faced, it ceases to be a force for positive change in a complex and contested world. To create more sustainable societies requires significant changes in the way we live, organize our communities, and do business. And change, especially radical change, will always have its opponents. As such, advocates of sustainability must be wary of superficial consensus.[6] If we are so eager for sustainability to be universally endorsed that we are unwilling to offend or oppose anyone in its pursuit, then we will achieve nothing of merit. The pursuit of sustainability need not encourage conflict; but it cannot avoid it. If sustainability becomes a "feel good" issue that shies away from all controversy in pursuit of consensus, then a crucial battle will already have been lost. Pursuing sustainability requires taking a stand.

Allies of sustainability are, oftentimes, fair-weather fans without deep commitments. We can expect many who endorse it by name to shirk it in practice. The same could be said, of course, for most any ideal. This is not to recommend a cynical perspective. Cynicism forgoes the hard work of having to determine, in each case, when hypocrisy or deceitful advertising is in play, when compromises go too far, when the practice employed to achieve a valued goal betrays its principles, and when opponents must be squarely confronted. And cynicism forgoes the crucial task of determining how the inevitably flawed means we employ to pursue our ideals might be improved. Sustainability is a twisting path blazed through rough terrain. It is best pursued when informed by principle, steadied by hope, directed by practical judgment, and is welcoming of all potential allies. It cannot afford cynicism. There is too much at stake.

A pragmatic ideal

Sustainability cannot afford naive idealism. It needs to be more than an endorsement of a better world. It needs to be a practice. To speak of *practicing* sustainability underlines two important points. First, sustainability is an activity. Second, sustainability is something we pursue but never perfect. We *practice* sustainability in the same sense that we *practice* a musical instrument. There is always room for improvement. That is the point of practicing. And as the world is ever changing, the balance that sustainability seeks to establish between ecological, social, and economic welfare is dynamic, not static. To practice sustainability is to promote adaptive change. That is why sustainability can only be pursued by way of cultural creativity grounded in learning. Sustainability is not a destination to be reached, but a winding path to be well traveled. Practicing sustainability entails life-long learning.

Ideals are things to be reached for but are rarely fully grasped. But sustainability is different from many other ideals, such as liberty. If you only care about a single good, such as personal freedom, then it is consistent and theoretically possible to seek its full measure. Libertarianism, for instance, is an ideology that celebrates the value of individual liberty and recommends its maximization in all situations at all times, whatever the results. In contrast, sustainability is not a single value. It entails the balanced pursuit of at least four distinct goods.

If you like apples and do not like broccoli, rice, or carrots, then you can indulge your preference for this fruit and never consume these other foods. But if you like apples *and* broccoli *and* rice *and* carrots, then you cannot maximize all four preferences. Every time that you eat an apple, you could be eating broccoli or rice or carrots, and vice versa. It is impossible simultaneously to pursue each of a number of diverse goods in full measure. Sustainability is not the effort to maximize a singular good. It requires us to combine, integrate, and balance ecological health, economic welfare, social empowerment, and cultural creativity. These four goods are compatible and mutually supportive. But that does not mean that all of these goods can be simultaneously maximized. Each and every day, compromises have to be made.

The environmental group Earth First! proudly bears a slogan on its journal's masthead: *No Compromise in Defense of Mother Earth*.[7] Over the four decades of its existence, Earth First! has defended forests, watersheds, and ecosystems from resource extractors and developers, oftentimes by destroying their machines. Earth First!'s methods are radical, some would say extreme. Then again, so is the greed and destructiveness of many of their targeted opponents. But the environmental organization unnecessarily gives compromise a bad name.

Notwithstanding Earth First!'s slogan, no one who partakes of modern civilization and industrial society can escape compromising in his or her effort to promote ecological health. To buy durable goods from a store, drive a vehicle or fly in a plane, heat or cool your home, or consume food means that you have deprived the planet, an ecosystem, or another species of a valuable resource, and often a life. We compromise in our defense of Mother Earth the moment we get out of bed in the morning and step into a hot shower or eat breakfast. And it does not obviate the compromise, though it may greatly diminish it, if our water is heated by solar energy and our breakfast made from organic foods. However sustainably we live, it is virtually impossible to avoid depleting resources and adversely impacting the welfare of other life forms.

If sustainability were solely a matter of minimizing our impact on the planet, then we should consume the least we can and never propagate. Indeed, if minimizing impact is the goal, then the best thing is not to have been born in the first place. But sustainability is not simply a matter of minimizing our negative impacts. It entails maximizing our positive impacts. The point is not to erase our existence, but creatively to embrace our responsibilities.

Sustainability is a pragmatic affair. Because it entails pursuing multiple goods in tandem, its advocates ought to wear their willingness to compromise as a badge of honor. Absent this balancing act, sustainability would justify the accusation of being a "gigantic exercise in self-deception."[8] The fact that compromise is unavoidable should not dishearten us. Compromise, at its best, means achieving a healthy balance. To eat a balanced diet is to recognize that there is no one food that yields optimal health. Eating apples, broccoli, rice, *and* carrots provides needed variety in one's diet. Likewise, ecological health, economic welfare, social empowerment, and cultural creativity mix very well together and, in the right proportions, prove mutually supportive. The effort to maximize any one of them in isolation, in contrast, will undermine the chances of achieving a good measure of all.

Sustainability is both a powerful ideal and a pragmatic mandate. It speaks to our lives as citizens and leaders, neighbors and community members, consumers and business people, students and teachers, workers and professionals. It affects what we buy and sell, consume and discard, how we travel and what we eat, whom we vote for and how we interact socially. It bears moral weight, political implications, and cultural consequences.

At times, sustainability appears all-encompassing in its scope. But should sustainability subsume all of our values and aspirations? William McDonough and Michael Braungart observe that "If a man characterized his relationship with his wife as sustainable, you might well pity them both."[9] Why is sustainability seen as an ideal for society, but a pretty meager achievement for personal relationships? Given that almost half of all first marriages today end in divorce within 15 years, a sustainable romantic partnership is perhaps quite an achievement. But few of us would say that the notion of simply *sustaining* a relationship, in the sense of maintaining it over time, excites our passions. We want *joy*, *intimacy*, *spontaneity*, and *fulfillment* from love and from life. None of those words rhymes with *sustainability*.

Should they rhyme? Should sustainability be understood as something optimal and sufficient for our lives? Does sustainability trump all other ideals? Or is it simply a means that allows us to pursue higher goals? If so, how do we balance its practice with the pursuit of other cherished goods? In the following pages, we will grapple with such questions.

Chapter 1 engages the history of sustainability as a concept and practice. More broadly, it examines the relationship of sustainability to expanding time horizons that relate us both to ancestral legacies and future generations.

Chapter 2 investigates our spatial horizons, the global geography of sustainability and its social foundations. Duties to future generations are compared to the obligations we have to contemporaries who share our communities and planet.

Chapter 3 explores the ecological foundations of sustainability. We examine the value of biodiversity and environmental health, and the nature of resilient ecosystems.

Chapter 4 probes the impact of technology. It underlines the importance of green innovations while confronting the side effects of technological solutions.

Chapter 5 addresses political and legal issues. We investigate the role of political parties, policy and regulation, law and the courts, as well as the changing nature of national security and forms of governance.

Chapter 6 examines the economic foundations of sustainability. It tackles limits to growth in relation to current levels of consumption and waste, while exploring the demands and opportunities of an ecological economics.

Chapter 7 explores the cultural foundations of sustainability, including its intellectual, aesthetic, and ethical dimensions. It considers the role of spiritual traditions and education in fostering cultures of sustainability characterized by both creativity and the conservation of core values and relationships.

The Conclusion challenges us to confront the prospect of the collapse of civilization and to embrace sustainability as a journey that can inspire us as individuals, ground us in community, and help us redefine prosperity.

Inquire and Explore

1.   Is sustainability an enduring worldview and practice, a "megatrend," or just a fad and buzzword?
2.   How rampant is greenwashing, and what can be done about it?
3.   Are you living sustainably if you are energy efficient and always recycle?

* * * * * *

Get your sustainability lifelong learning off to a great start!

* Browse a handy list of green facts and check out crucial global data and real-time statistics.
* Explore a planetary encyclopedia and test your eco-literacy.
* Hook up with key sources of sustainability news and podcasts.

 For web resources and more information on these topics, please visit conservationandcreativity.net

1
Sustainability and Time

Do we have the right to burden our descendants with economic debt, social disorder, resource depletion, ecological degradation, and increased insecurity? Is it legitimate for us to enjoy a high quality of life at their expense? What are our responsibilities to future generations? These are questions of intergenerational justice, a core component of sustainability.

This chapter explores the relationship of sustainability to time. It starts with the history of sustainability as a concept and a practice. In turn, it examines how sustainability expands our time horizons, providing a sense of responsibility toward future generations, as well as an appreciation of our inheritance from the past. Many of the virtues and skills that ground sustainability are part of this inheritance. While exploring this legacy, the chapter also surveys the future of sustainability as a set of evolving ideas and practices.

If current trends continue, future inhabitants of the earth will face: a hotter and more disruptive climate; rising, acidified oceans; vastly diminished biodiversity; and natural resources in short supply. It will also be a more crowded world, more polluted, and prone to greater social and political instability. Life will be more precarious. Imagine that your birth had been delayed by a few decades. How would you feel if you were to inherit this depleted and dangerous world?

Advocates of environmental protection appear uncompromising when they insist that nature must never be degraded for short-term gain. But they may not be as single-minded and anti-economic as they seem. Indeed, their efforts are often economically oriented and well balanced – for the long term. They seek to ensure that today's profits are not gained at the expense of future prosperity. An Oregon conservation director put the matter well when he assessed local prospects for logging, agriculture, and energy production. Insisting that these economic opportunities be kept in balance with the preservation of old-growth forest and the protection of salmon and other endangered species, he observed that "Environmentalists may be hell to live with, but they make great ancestors."[1]

These words apply well to advocates of sustainability. The welfare of future generations is a foremost concern. However, to make future generations a priority is not to determine how progeny will live. That would be paternalism. We would not want our ancestors to have fixed our fate. Our children and grandchildren deserve every bit as much autonomy and opportunity as we enjoy, ideally even more. So the task at hand is the "conservation of options."[2] The way we lead our lives and conduct our business today, whether we recognize it or not, will impact the prospects of our descendants. To practice sustainability is to live and work in a way that does not foreclose meaningful and abundant choices. We ought not to deprive future generations of the opportunity, for example, to experience vast areas of wilderness and high levels of biodiversity, a stable climate, and healthy, unpolluted environments.

Who could object to the conservation of options for future generations? The problem is that the future is a far-off place, and the present has so many demands and enticements. So we tend to discount the future. Al Gore – former vice president of the United States and Nobel Peace Prize winner for his efforts to combat climate change – summed up the problem. In Gore's words, "the future whispers while the present shouts."[3] Giving a more equal voice to the future is no easy task. Even in our own lives, we often find it difficult to forgo present enticements with an eye to the days, weeks, months, and years ahead. Accordingly, people often find themselves in debt, unprepared for examinations or professional tasks, and suffering from lack of sleep, excess of drink, or ill health. Present comforts and pleasures loom large, while distant consequences shrink from view. We are often myopic creatures.

At the same time, human beings naturally care about the future and, in particular, about the future of their descendants. Parents' concern for their children is legend. Not infrequently, mothers and fathers make extraordinary sacrifices for their sons and daughters, risking life and limb to secure their welfare. This capacity for selfless sacrifice, too, is a fundamental feature of our species.

But concern for future generations is not only a matter of making sacrifices. Children and grandchildren are a source of joy. Even beyond their direct descendants, people take great pride in leaving a legacy to the future citizens of their towns, cities, and nations. The philosopher Ernest Partridge observes that concern for the welfare of posterity allows us to identify with "larger, ongoing, and enduring processes, projects, institutions, and ideals." In the absence of this breadth of vision, Partridge writes, our lives would be "empty, bleak, pointless, and morally impoverished."[4] A future focus brings purpose to our sojourn on earth. It makes our lives fulfilling in ways that an existence wholly given over to satisfying immediate needs and wants cannot. Meeting our obligations to future generations produces appreciative descendants. But fully experiencing this responsibility – and the meaningfulness that accompanies it – is a crucial good for us in the here and now.

The history of sustainability

There is an old saying that "Forests precede civilization; deserts follow." The collapse of ancient societies often followed the deforestation of their lands. In fact, a concern for deforestation in the 1700s led to the coining of the term *sustainable*. The smelting of ores to produce metals in Germany required large amounts of wood to fire the furnaces. German forests were so decimated by the overharvesting of timber that the mining industry was threatened. In response to the crisis, Hans Carl von Carlowitz, a mining administrator, wrote the first book on forest management in 1713. Here, he outlined methods for the *sustainable* use (*nachhaltende Nutzung*) of woodlands. Carlowitz argued that well-managed forests could supply timber resources indefinitely.

Notwithstanding Carlowitz's early use of the term, the vocabulary of sustainability did not find its way into common usage until the 1980s. But in the prior century, increasing concern for the welfare of forests and natural habitat foretold its eventual appearance. Having witnessed the destruction of the forests of the Green Mountains in his native Vermont, and similar deforestation in Europe, George Perkins Marsh strongly advocated for the restoration of degraded landscapes and watersheds. With the publication of *Man and Nature* in 1864, Marsh set the stage for the early conservation movement. He warned of the "dangers of imprudence and the necessity of caution" whenever humans "interfere with the spontaneous arrangements" of nature.

A generation later, Gifford Pinchot, the chief of the United States Forest Service in the late 1800s, developed a "conservation ethic." Implementing the lessons he learned while studying German forestry practices, Pinchot became known as the "father of conservation." He wrote:

> The central thing for which Conservation stands is to make this country the best possible place to live in, both for us and for our descendants. It stands against the waste of natural resources which cannot be renewed, such as coal and iron; it stands for the perpetuation of the resources which can be renewed, such as the food-producing soils and the forests; and most of all its stands for an equal opportunity for every American citizen to get his fair share of benefit from these resources, both now and hereafter.[5]

Pinchot's treatise on conservation, published in 1910, captures much of what we mean today by sustainability. His efforts led to the development of "conservationism," which became a popular approach to managing natural resources.

Pinchot's resource-conservation efforts were complemented by John Muir's struggle to preserve wilderness. Muir, an avid outdoorsman and the founder of the Sierra Club, befriended Pinchot and applauded his efforts. Eventually, however, they had a parting of ways. While Pinchot focused on conserving natural resources to ensure long-term human benefits, Muir wanted to safeguard the natural world from the degradation caused by its exploitation. Muir insisted that nature had intrinsic value, apart from its worth as a storehouse of resources for human consumption. There was an aesthetic and spiritual renewal to be gained from the experience of nature. For Muir, satisfying basic human needs through the consumption of natural resources was unavoidable. The question was whether we, and future generations, would also be able to satisfy our recreational, aesthetic, and spiritual needs for unpolluted landscapes, open spaces, scenic beauty, diverse wildlife, and untamed wilderness. Muir's efforts led to the development of "preservationism." Preserving nature required a reverential attitude, a belief in nature's inherent value.

Aldo Leopold worked in the Forest Service under Pinchot's supervision. Like his boss, Leopold was an advocate for the scientific management of natural resources, including wild-life, to maximize the long-term benefits for people. At the time, that meant killing wolves and other predators to mini-mize losses of livestock and to increase deer herds for hunters. The extermination program was a huge success. But it proved a Pyrrhic victory. Leopold sadly witnessed deer populations, absent natural predators, overshoot the carrying capacity of the land. Having denuded their habitat of its foliage, the quickly expanding deer herds then suffered mass starvation.

Following the conservation model of Pinchot, Leopold aimed to secure the long-term benefits for the greatest number of citizens. But these early efforts at resource management were carried out with insufficient ecological understanding. Consequently, they began to unravel the web of life. Faced with a disaster of his own making, Leopold came to under-stand that humankind must transform itself from "conqueror of the land-community to plain member and citizen of it."[6] To aid in this transformation, Leopold proposed "thinking like a mountain."

Thinking like a mountain means seeing ourselves as parts of a complex, interconnected, ecological web of relations, rather than entitled owners of resources available for endless exploitation and consumption. Effectively, Leopold blended Pinchot's concern for the "wise use" of natural resources with Muir's sensitivity to the wonders of nature. In turn, he added his own insights regarding the complexity and connectivity of ecological systems.

To understand ecological systems, we have to understand their evolutionary development. "The last word in ignorance is the man who says of an animal or plant: 'What good is it?'" Leopold wrote. "If the biota, *in the course of eons*, has built something we like but do not understand, then who but a fool would discard seemingly useless parts? To keep every cog and wheel is the first precaution of intelligent tinkering."[7] Evolution has created a complex system of interdependent parts. Leopold emphasized the importance of preserving all the parts and considering each in terms of its contribution to the whole.

Most of the key figures that set the stage for sustainability over the next century, like Leopold, fell somewhere along the spectrum that runs between Pinchot's concern for the fair and frugal exploitation of natural resources and Muir's concern for the intrinsic worth of nature and its non-consumptive (recreational, aesthetic, and spiritual) benefits. But developments in technology and rising human populations would push these concerns into new arenas.

Rachel Carson's 1962 best-seller, *Silent Spring*, investigated the devastating impact of pesticides and other industrial chemicals. Paul Ehrlich's renowned 1968 book *The Population Bomb* confronted the general public with the dire consequences of rising human numbers. Barry Commoner's widely acclaimed *The Closing Circle*, published in 1971, identified technology and the inequitable distribution of resources as the primary causes of human suffering and environmental destruction.

Bringing many of these concerns together, *The Limits to Growth* examined the impact of increasing human populations, pollution, and natural resource consumption. The publication of this book in 1972 was a watershed event. For the first time the goal of sustainability was applied not to isolated practices, such as forest management, but to the global system. The challenge was to model a world that was "1. *Sustainable* without sudden and uncontrollable collapse; and 2. capable of satisfying the basic material requirements of all of its people."[8] The conclusion reached by Donella Meadows and her co-authors was chilling: "If the present growth trends in world population, industrialization, pollution, food production, and resource depletion continue unchanged, the limits to growth on this planet will be reached sometime within the next one hundred years. The most probable result will be a rather sudden and uncontrollable decline in both population and industrial capacity."[9] With 30 million copies distributed in 30 languages, *The Limits to Growth* became the best-selling environmental book ever written.

Critics responded sharply. Optimists with faith in the human capacity to devise technological solutions for every problem either denied that there were limits to growth or maintained that these limits would not be reached in the foreseeable future. Julian Simon argued that rising human populations simply means more people available to solve problems and develop technology. With this in mind, Simon asserted that "natural resources are not finite," and that growing populations could indefinitely secure and maintain ever-rising standards of living.[10] Likewise, Herman Kahn insisted that in the next two centuries people "almost everywhere... will be numerous, rich, and in control of the forces of nature."[11]

In a 30-year update to *Limits to Growth*, the authors responded to their critics. Acknowledging that a number of specific predictions proved inaccurate, they argued that their prognosis had largely been vindicated. An accurate assessment of the current state of the world, they observed, demonstrates that global levels of pollution have not leveled off or declined and that technology has not allowed us to substitute newly discovered or invented substances for most of the earth's declining stocks of natural resources.

To be sure, substitutes for particular resources have been developed, and in particular countries or regions, pollution has been significantly reduced. But, from a global perspective, natural resources continue to decline precipitously and pollution is increasing at an alarming rate. The earth's capacity to serve as a stock of natural resources (such as oil, fertile land, and fresh water) and a sink for pollution (such as greenhouse gases) is reaching or has already surpassed its limits.[12]

With similar concerns, the International Union for the Conservation of Nature and Natural Resources (IUCN), formerly called the World Conservation Union, addressed the need to manage "air, water, soils, minerals and living species including man, so as to achieve the highest sustainable quality of life." In 1980, the terms "sustainable utilization" and "sustainable development" were extensively employed in the IUCN's *World Conservation Strategy*. This Report, subtitled *Living Resource Conservation for Sustainable Development*, was the first publication to employ the term "sustainability." It did so in reference to the goals of conservation and with explicit mention of intergenerational obligations.[13]

The following year, Lester Brown published *Building a Sustainable Society*. Brown had founded the Worldwatch Institute half a dozen years earlier, the first research organization devoted to the study of global environmental issues. Brown's book began with a reflection on the ancient Mayans. The Mayans lived in Central America for almost three thousand years, steadily growing in population and agricultural productivity. Then, quite abruptly around 900 CE, a number of Mayan centers collapsed. The causes are not known with certainty, but deforestation and the degradation of cropland appear to have played major roles. Brown goes on to grapple with some of the toughest issues facing the world today, such as threats to food production, degraded soils, deforestation, overpopulation, loss of biodiversity, and the need for renewable forms of energy.

Building a Sustainable Society outlined the direness of the challenges we face at a global scale. It questioned whether we might soon suffer the same fate as the ancient Mayans. In outlining prospects, however, Brown concluded with a sense of optimism:

> Creating a sustainable society will require fundamental economic and social changes.... The magnitude of these changes is scarcely in question. Every facet of human existence – diet, employment, leisure, values, politics, and habits – will be touched. As the transition proceeds, new skills will be needed and old skills will become obsolete.... In effect, we have embarked on a shared adventure, the building of a society that has the potential to be an enduring one. This awareness could begin to permeate almost everything we do, imbuing it with a sense of excitement – one that derives in part from the scale of the undertaking, which has no precedent, as well as from full knowledge of the risks and consequences of failure.[14]

Brown's annual *State of the World* publications, published by the Worldwatch Institute, began charting "progress toward a sustainable society" in 1984.

In 1987, the World Commission on Environment and Development, sponsored by the United Nations, published the Brundtland Report. The most popular definition of sustainability comes from the pages of this report, named after the chairperson of the Commission, the Norwegian prime minister Gro Harlem Brundtland. Here, development is considered sustainable if it "meets the needs of the present without compromising the ability of future generations to meet their own needs."[15] The report encouraged economic development that improves the lives of the world's poor without diminishing the prospects of future generations.

In the century preceding the World Commission, sustainability was an emergent concept lacking full articulation. With the publication of the Brundtland Report, sustainability found its linguistic feet and quickly moved into the mainstream of governmental and scholarly interest. Since the early 1990s, sustainability has been considered the "dominant global discourse of ecological concern."[16] It was a centerpiece of the 1992 United Nations Conference on Environment and Development, the so-called "Earth Summit" held in Rio de Janeiro. It gained a global audience again in the 2002 World Summit on Sustainable Development in Johannesburg, South Africa, at the "Rio + 20" Earth Summit in 2012, also held in Rio de Janeiro, and in scores of other regional and international conferences.

In the last quarter-century, sustainability has become the focus of thousands of scholarly articles and books, as well as hundreds of academic programs, governmental agencies, and citizen-based organizations. During this time, many former "environmental" organizations concerned solely with nature preservation came to acknowledge sustainability as the "driving force" of their efforts: "Today we recognize that minimizing pollution and resource damage is not good enough," observed the head of the Audubon Society, an environmental group primarily concerned with the conservation of wild birds and their habitat. "The real objective must be to meet human needs – both physical and spiritual – without limiting opportunity for future generations."[17]

To summarize, the sustainability movement grew out of the conservation and preservation movements of the nineteenth century and the environmental movement of the twentieth century. Its most important and original contribution to these earlier movements is an insistence that the protection of nature is most effectively achieved when economic development and social empowerment are simultaneously pursued. Definitions of sustainability continue to evolve. Approaches to its practice by local, national, and international organizations and programs vary widely. Still, a core feature remains the effort to balance the satisfaction of current needs with the safeguarding of future prospects.

Expanding time horizons

In its seminal description and endorsement of sustainability, the World Conservation Union observed – in words often repeated – that "we have not inherited the earth from our parents, we have borrowed it from our children."[18] This popular saying offers a trenchant insight. The scant days that we spend on earth are dwarfed by the eons that belong to our descendants. So while our lives are often focused on immediate needs and wants, moral duty demands a future focus.

Catchy sayings often pack a punch. But they can misrepresent the facts. To be sure, we *are* borrowing the earth from our descendants. At the same time, our parents and grandparents have passed the world down to us. We *do* inherit the earth from ancestors, generations of forebears dating back millennia. The most common definitions of sustainability address our responsibilities to future generations. But to fully grasp the meaning of sustainability, we have to explore our relationship to the past as well.

Edmund Burke, the eighteenth-century British political philosopher, observed that human society is "a partnership not only between those who are living, but between those who are living, those who are dead, and those who are to be born."[19] Burke's notion of a partnership across time accords well with the concept of sustainability. The idea, as Burke noted, is that current generations ought not to waste their inheritance. They should live with "what is due to their posterity" well in mind.[20] To live sustainably is to think and act with an expanded time horizon that puts one in partnership with ancestors *and* progeny.

Imagine yourself standing in the present moment. As you gaze ahead at your descendants, do not stop at one or two generations. Venture further, as the effects of your actions will not limit themselves to a few score years. Your deeds will impact the lives of great-great-great-grandchildren. Consider the fact that the fossil fuels we burn today will likely have their most drastic effects on climate change in the coming centuries. The radioactive waste that we produce in nuclear reactors will remain toxic for at least 10,000 years. Indeed, the effects of our actions extend even further. The non-renewable resources that we deplete will be unavailable for *all* future generations. And when we cause the loss of biodiversity by our destruction of habitat or overharvesting of a species, that loss will be suffered by *every* future inhabitant of this planet. Extinction is forever.

Contemplating the distant future stimulates a sense of responsibility. It is equally sobering to reflect upon the distant past. Most of our inheritance was not bequeathed to us by our parents or grandparents. It is the gift of thousands of years of cultural development, millions of years of biological development, and eons of geologic development. The benefits of social and political life and technological innovation have been steadily accruing ever since our species left its Stone Age caves. Our ecological inheritance – earth's supportive web of life composed of countless species – has been in development for tens of millions of years. The same can be said for many of our renewable resources, including fresh water, fertile soils, and benign atmospheric conditions. And the non-renewable natural resources that we extract from the earth's crust to power industrial society were formed and deposited hundreds of millions or billions of years ago. Thinking about sustainability situates us in deep time. Deep time refers to the countless millennia that have already unfolded to bring our particular lives into being, and the countless millennia yet to come that will be profoundly shaped by what we do, or fail to do, with these lives.

A consideration of our species' relationship to carbon provides a powerful way to experience deep time. Carbon, in various forms, is stored in the earth's biota, in its soils, and in its geologic deposits. Had it stayed in these dispersed storehouses, our species might never have moved beyond a hunter-gatherer existence. Civilization developed because our species figured out how to extract and exploit carbon ever more efficiently.

As agriculturalist Wes Jackson observes, 10,000 years ago the first trappings of civilization arose through the agricultural surplus produced from carbon-rich soils.[21] And 5,000 years later, the carbon stored in trees was burned to smelt bronze and then iron, transforming primitive agricultural communities into kingdoms with tools and weapons. After another 5,000 years, the carbon stored in coal was exploited to fuel the Industrial Revolution. In the following century, even more energy-dense forms of carbon – such as petroleum and natural gas – were pumped out of the earth to produce the fossil-fuel economies we have today.

Abundant energy-dense forms of carbon have allowed us to harvest ever more resources from the earth at an accelerating rate. Soils produce high yields of grains, fruits, and vegetables by being infused with fertilizer made primarily from natural gas. Machinery fueled by gasoline and diesel allows us to harvest these foods, and to extract and process tremendous amounts of other natural resources, such as fish, timber, and minerals. Our built environment is largely powered and maintained by fossil fuels, and the lion's share of the durable goods that fill our homes and businesses, including most plastics, are manufactured from petrochemicals.

Petroleum has nearly twice the energy density of coal, which has twice the energy density of wood. A single barrel of oil yields about the same amount of energy as would be produced by eight years of a sturdy person's physical labor. So the exploitation of carbon has vastly increased material productivity and, consequently, the human population. Indeed, our species' numbers have risen at roughly the same rate as more energy-dense fuels have been burned, increasing tenfold over the past three centuries and fourfold in the last hundred years.

Each calorie of food that we consume today, on average, requires about ten calories of fossil-fuel energy to arrive on our plates.[22] A human being needs about 3,000 calories a day for nutritional health. The average US consumer expends 60 times that amount of energy each and every day.

Within the blink of an eye in the scale of geologic time, we have managed to exploit much of the earth's stored carbon. As a result, our species has grown exponentially in its numbers while fundamentally transforming the land, oceans, and skies of the planet. Since the dawn of the Industrial Revolution, we have put half a trillion tons of carbon into the atmosphere, and we will deposit another half trillion tons in the next 40 years if current trends continue.

Civilization's 10,000-year history with carbon gives us a sense of deep time. That sensibility can be enriched by a journey that takes us across the full expanse of planetary history. We begin when the earth was formed billions of years ago from the remnants of an exploding star. Rather than imagining billions of years, however, we will condense this massive swath of time into a span easier to conceive: a single year.[23] Assuming, then, that the planet was born at midnight on January 1, each of the following 365 days will represent 12.4 million years of the earth's 4.54 billion-year history.

The extreme conditions on the young, hot planet forbade the development of life for the first few months after its birth. Only in March did single-celled organisms develop in the cooling seas. Meanwhile, a protective atmosphere that included a layer of ozone was forming, shielding the planet from searing solar radiation. This allowed the first bacteria and blue-green algae to evolve. By July, multicelled organisms had developed. But life on the young earth was still moving at a protozoan pace. It was not until late November that complex organisms appeared on dry land. First, there were plants, then insects and spiders. Only in December, more than 11 months into our year, do we find reptiles crawling about. Within a week, however, these scaly creatures have grown to a tremendous size. The reign of the dinosaurs begins.

Meanwhile, the planet's continents have been forming, drifting apart, and reforming. The last supercontinent to congeal, Pangaea, begins to break apart to form our current seven continents at the end of the second week of December. Within a few more days, the earliest birds and mammals appear. By the end of the third week of December, the dinosaurs are already extinct. Mammals now grow in number, size, and diversity.

Then, in the early morning hours of December 31, one particular mammal, the first hominin known as *Australopithecus afarensis*, begins to walk upright. With an hour to spare before midnight, our Stone Age ancestors learn to control fire. Eight minutes before the year comes to an end, *Homo sapiens*, the contemporary human race, arrives on the scene. At 11.59 p.m., our species begins its agricultural existence, having learned to domesticate plants and animals. Christopher Columbus sails across the Atlantic Ocean to initiate the process of modern globalization at four seconds before midnight. There are about 500 million humans now occupying the planet. Before the end of the day, that number will increase fourteen-fold.

With only two seconds left, the Industrial Revolution sets the world on a course of rapid transformation, exploiting carbon deposits that first formed underneath the earth's crust during early December, in the Paleozoic era. At the hands of a quickly rising population, agricultural, technological, and economic developments now impact life on every continent, fundamentally altering atmospheric conditions and producing a crisis of extinction. At a quarter of a second before midnight, visionary individuals first articulate the concept of sustainability, and work feverishly to ensure that their species and its planetary home will see a new day.

With global transformation accelerating at an exponential rate, we cannot predict the fate of our species even through the first second of New Year's Day. But, most assuredly, as a result of climate change produced by the burning of fossil fuels and destruction of forests, the planet's oceans will rise by a meter or more, washing away thousands of islands and shrinking coastlines on every continent. Owing to rising seas and disrupted climates, hundreds of millions of people will be forced to migrate, global agriculture will be thrown into disarray, and civilization itself will face an unprecedented threat. There is no certainty that *Homo sapiens*, and countless other species that now grace the planet, will survive the first day of the new year.

We do know that the earth itself will not endure through a full second year. Our planet will eventually be consumed by the sun, which is itself undergoing constant transformation.

Even now, the sun is steadily moving toward the latter stage in the life of a star, when it becomes a red giant. In time, its radius will increase 200-fold, causing it to engulf the first three planets of the solar system. Well before the earth is transformed into molten rock, however – indeed as soon as March of its second year – the oceans will start to evaporate. This will signal the beginning of the end of earth's 5-billion-year experiment with life.

To stand back and contemplate planetary history puts us in a state of awe. We marvel at the expanse of cosmological eons and geological eras, and at the scope and power of the forces at play in our universe. Awe is closely related to fear, and a voyage into deep time can produce anxiety. How can we measure ourselves – our capacities and aspirations – against such vast temporal horizons? What possible impact can we have as individuals, with our measly four-score years of life, or even as a species, with our few million years of existence, when we set ourselves against the billions of years that define the planet's odyssey?

Notwithstanding what you may hear in the popular press, sustainability is not about "saving the planet." We did not create earth and cannot destroy it. Indeed, life on earth has undergone many catastrophes and has always, in time, bounced back. At least five mass extinctions have occurred on the planet. More than 99 percent of all the species that have ever existed are now gone forever. But here's the rub: we are currently in the midst of a sixth mass extinction, and unlike the previous five, this one is the work of our own hands. Likewise, while the planet has undergone tremendous climatic variation over its history, current global warming is occurring at an unprecedented rate, threatening civilization. And we are its cause.

Viewing sustainability through the lens of deep time gives us a humbling sense of perspective, helping us find our place on a very big planet – and a much larger universe – across a mind-boggling expanse of time. To embrace sustainability is to adopt the perspective of the two-faced Roman god of portals, Janus, with one set of eyes looking forward and another set looking backwards. Situated historically in the context of eons of planetary and cosmological development, we can appreciate how small a part we play.

The question is: how well will we play our part? Becoming sensitized to deep time need not disable us from skillfully and responsibly acting in the present. Quite the opposite. By fostering a sense of gratitude for our inheritance, responsibility for our legacy, and both humility and pride for the innovative roles we can play within a vast cosmological saga, an expanded time horizon can foster commitment to caretaking and creativity.

Ancient virtues and enduring challenges

Though the word *sustainability* found its way into common parlance only in the last few decades, the concept behind the word is deeply indebted to the distant past. Generations of humans over thousands of years have cultivated the knowledge, skills, and virtues that set the stage for its appearance. To fully understand the meaning of sustainability and its contemporary importance, we have to appreciate its historical vintage.

Sustainability saw its genesis in ancient agrarian societies. Prior to the development of agriculture some 10,000 years ago, humans existed in small hunter-gatherer tribes. They gathered fruits, seeds, and roots, and hunted wildlife. If they depleted key resources in a particular area, they moved on to live off the fat of the land – or scrape by – elsewhere. Early hunter-gatherers likely had a reverence for the plants and animals that sustained them. And it is very likely that they frowned upon waste, for they undoubtedly knew hunger all too well. But their populations were small, the earth was large, and the impacts of their actions were quite limited.

With the advent of agriculture, everything changed. Surplus from harvested crops could support larger populations. These populations were less migratory. They stayed on the land that produced their food. No longer nomadic as their hunter-gatherer forebears, agricultural societies built permanent housing, acquired more extensive chattels, stored grain, and raised livestock. We have no written records of the earliest agricultural societies. But we might guess how something like the concept of sustainability arose within these communities.

Imagine a scene portraying the hardships that early agrarian communities undoubtedly faced, and the knowledge and virtues they developed to survive:

The March snows lay heavy and deep across the open fields to the east, land that had betrayed no sign of life for six months now. It is very cold, and the whitened forest that surrounds the village and its adjacent farmland seems to have lost all of its power to curb the bitter winds. They howl and bite at will. Everyone huddles indoors. The mood is gloomy.

Provisions duly gathered and stocked before the first snows fell – now a full half-year ago – are nearly exhausted. There is plenty of wood for the fires that warm the beleaguered inhabitants of the sod huts. But the storehouses of grain are almost empty, and with snowdrifts high and temperatures low, success in hunting and foraging has been rare. Rations are skimpy. People are forced to eat food that is spoiling. Young children cry with hunger before sleep finally stills their voices.

Villagers gather to discuss their dire state in the large roundhouse built a decade back, the pride of the community. A suggestion is made by a young father of four children to open up the caches of seed corn. These underground storehouses hold enough grain to feed everyone for two weeks. By then, he argues, the snow will be gone and hunting parties can venture further to find prey. Soon enough, the green earth will show its face again, tubers can be dug and fresh shoots gathered.

The thought of steaming porridge widens the wane eyes surrounding the fire that burns in the center of the log house. But the village elders quickly break the spell with harsh reprimands. The thought of weeks of meager rations and hungry children gnaws at their hearts, too. But without seed corn to be planted in the spring, the elders know, there will be no harvest in the fall. And without a harvest of grains to store, the following winter would be their people's last. The village would perish. Survival depends upon sacrifice. The elders speak with one voice: "Never eat your seed corn."

Such scenes undoubtedly played themselves out in countless villages across the continents over the millennia. Ever since human beings invented agriculture, people have stored grains over the winter for spring planting. "Never eat your seed corn!" probably became a watchword of agricultural communities.

Village elders were undoubtedly the spokespeople for sustainability. They passed on crucial skills and knowledge by mentoring youth. Technological innovation was rare. Slowness of change led to a heightened appreciation of inheritance. Ancestors and elders were revered, and often worshipped. Current generations understood themselves as trustees of a culture handed down by venerated forebears.

Greek youth in the ancient city of Athens, for instance, were required to attend the Ephebic College to gain the status of citizens. Graduating youth would pledge to maintain the ideals and revere the laws of the city, never to disgrace it by immoral acts, and to cultivate a spirit of civic duty. The Ephebic Oath concluded with the words: "Thus in all these ways we will transmit this City, not only not less, but greater and more beautiful than it was transmitted to us." The Ephebic Oath is a 2,500-year old declaration of sustainability.

Sustainability demands that actions taken today not jeopardize tomorrow. As the Ephebic Oath underlines, however, the knowledge, virtues, and skills that allowed ancient societies to live sustainably were expected to deliver more than basic survival. These qualities were cultivated not merely to safeguard life, but to enable the good life. Passing on the cultural legacies that foster the good life was a sacred trust.

The ancient practice of sustainability entailed the cultivation of specific virtues, such as frugality, simplicity, balance, moderation, and prudence. The virtue of frugality refers to a simple, economical way of living that is not, at the same time, miserly. The Latin root of the word *fruges* refers to fruit. Frugality is a kind of economic fruitfulness, a bearing of fruit that proves sufficient because it is not wasteful. The virtue is grounded in a belief that resources are finite and is well captured by the aphorism "Waste not, want not."

Sumptuary laws that forbade extravagance were in place in many ancient societies, including China, Japan, and many Islamic nations. In the Roman Republic, sumptuary laws restrained the use and consumption of luxurious dress and food. The censors of Rome published a list of those guilty of living too lavishly, as a means of shaming them into more frugal lifestyles. Such laws were not solely concerned with conserving resources and limiting waste. They were also aimed at preventing commoners and merchants from imitating the appearance of aristocrats, thus maintaining traditional social status and hierarchies. In turn, they restrained the outward display of gross inequalities of wealth, which was believed to have corrupting effects on society.

The frugality cultivated in ancient Rome and in the Muslim world was similar to the virtue of *simplicity* celebrated by Taoists in ancient Asian societies. Both frugality and simplicity are akin to moderation. Moderation is the rejection of extremes. It requires us to keep things in balance. The ancient Chinese sage Confucius (551–479 BCE) explicitly espoused the virtue of restraint and embraced the "Doctrine of the Mean," which counseled equilibrium and balance. At about the same time, Siddhartha Gautama, often known as the Buddha, advocated the "Middle Way," a path of moderation that navigated between the extremes of sensual indulgence and ascetic self-denial. Likewise, in the *Tao Te Ching*, Lao-tzu observes: "Fill your bowl to the brim, and it will spill. Keep sharpening your knife, and it will blunt. Chase after money and security, and your heart will never unclench."[24] For the ancient Taoists, moderation or balance was key to achieving serenity and skillfulness in the art of life.

Moderation, also known as temperance, was a much-celebrated virtue in ancient Greece. The temple of Apollo at Delphi bore the inscription "Nothing in excess." This dictum was taken to endorse moderation as an individual and a social virtue. The ancient Greek philosopher Aristotle believed that all virtues incorporated moderation. Every virtue was a mean, that is to say, an appropriate balance struck between two extremes. To be courageous, for example, is to strike a mean between actions that are cowardly, which is to say overly fearful of risk, and actions that are foolhardy and needlessly place oneself at risk. Moderation was a pivotal virtue. Pursuing the "golden mean" was the path to the good life.

The ancient Greeks and Romans held prudence or practical wisdom to be another of the four cardinal virtues, complementing moderation, courage, and justice. To be prudent is to have foresight, the ability to imagine various possible futures arising out of current actions and conditions. This foresight allows one to proceed in moderate fashion to achieve the best possible results. Prudence demands an expanded time horizon. It builds on past experience and takes advantage of present opportunities to ensure future welfare.

Aristotle insisted that practical wisdom allowed for the development of all the other virtues. Every virtue is a mean between excess and deficiency. Practical wisdom helps one discern this mean. But practical wisdom is not simply a form of knowledge. Rather, it is a form of reason integrated with "correct desire."[25] In this sense, practical wisdom is a moral virtue, blending knowledge with well-directed passion to produce ethical action.

Marcus Tullius Cicero, the Roman orator and statesman, deemed prudence the greatest virtue. The aphorism "An ounce of prevention is worth a pound of cure" probably finds its origins in Cicero's dictum that "Precaution is better than cure." Prudence dictates that it is better to be safe than sorry. While the impact of fortune can never be eliminated, prudence enables the pursuit of the good life within the context of a well-ordered society. It does so through concern for the consequences of action. Cicero's appreciation of prudence finds parallels in many ancient societies. In the *Analects* of Confucius, for example, we read that "He who gives no thought to difficulties in the future is sure to be beset by worries much closer at hand."[26]

Prior to the arrival of Europeans in North America, the Mohawk, Oneida, Onondaga, Cayuga, and Seneca tribes of the northeast formed a confederation. These "People of the Longhouse" came to be known as the Iroquois nation. They developed an oral Constitution, called the "Great Law of Peace," which was binding on all tribes. The Great Law stipulated that decision makers, the "mentors of the people," should always consider the welfare of progeny. These mentors were described as having skins with a thickness of "seven spans." The phrase was interpreted to mean that each of their decisions should attend to its impact on seven generations of Iroquois. "Look and listen for the welfare of the whole people," the Great Law stipulated, "and have always in view not only the present but also the coming generations, even those whose faces are yet beneath the surface of the ground

– the unborn of the future Nation."[27] A number of the Founding Fathers of the United States, including Benjamin Franklin and James Madison, studied the Great Law of the Iroquois. It likely influenced their efforts to craft a Constitution that would endure over time and foster the long-term welfare of a young nation.

Ancient and indigenous societies found it necessary to cultivate the virtues of frugality, simplicity, balance, moderation, and prudence because they faced enduring threats to their survival and welfare. They established cultures of sustainability as a means of self-preservation. But they were not always successful in their efforts. Many ancient societies depleted their resources to the point of self-destruction.

Early agricultural peoples often overworked and over-irrigated their fields. The resulting erosion, siltation, and salinization of croplands contributed heavily to the downfall of civilizations. This was likely the fate of the ancient Sumerians in Mesopotamia, the Anasazi of southwestern North America, and the Maya of Central America. Ancient peoples armed with simple iron, bronze, or even stone axes often deforested their lands and depleted its wildlife. At times, the deforestation and depletion was so severe as to cause the collapse of societies. Such was likely the fate of the inhabitants of Easter Island. And many populations simply grew too large to support themselves. While the causes of the ultimate demise of ancient societies are complex, overpopulation and resource depletion appear to have played decisive roles.[28]

The future of sustainability

Time is not what it used to be. For the vast share of our species' history, people could count on one hand the number of technological innovations developed in their lifetimes. Social transformations were equally rare. That is not to say that life in days gone by was a static affair. It was likely full of adventure and learning. But most observable changes were linked to life cycles, as individuals moved from childhood through adolescence to mid-life and into old age. They experienced birth and death, health and sickness, periods of relative prosperity and famine, peace and war, and cyclical patterns of demographic growth and decline. But technological advances were few and far between, and changes in social structures, institutions, and customs were sparse. Sons' lives were modeled on their fathers' and grandfathers'; a daughter's world was virtually identical to her mother's and grandmother's.

Now technological and social developments are numerous, frequent, and accelerating. It is often said that the only thing you can count on in this life is change. Today, even that cannot be counted on. Change itself is changing. It is becoming more frequent. And it is getting faster. Most of the mid-career jobs of students reading this book will involve services, processes, and products that have not yet been invented. The best predictor of the future has always been the past. But in a world of accelerating change, our predictive capacities – even those well informed by history – are greatly challenged.

In his assessment of military preparations, Donald Rumsfeld, the US secretary of defense, said that "There are things we know that we know. There are known unknowns. That is to say there are things that we now know we don't know. But there are also unknown unknowns. There are things we don't know we don't know."[29] The past is filled with known knowns. It also harbors many known unknowns. The future is filled with unknown unknowns. This means that we do not even understand what sort of questions should be asked about the future in order to prepare ourselves for it.

Faced with a world of accelerating transformation and a future rife with unknown unknowns, how should we understand the quest for sustainability? A reasonable approach is to conceive it as the management of the scale and speed of change. Societies are like ecosystems. They survive by adequately responding to change. An ecosystem that suffers too sudden and too extensive a change will collapse. The same is true for a society. Managing well the scale and speed of change is central to sustainability.

Ecosystems are not wholly stable places. They are inherently erratic, variably respond to disruption, and have characteristics that shift in time. Incessant small-scale change allows long-term stability.[30] Adaptive change in the service of stability is called *resilience*. To be resilient is to withstand stress without breaking. An elastic band is resilient; it can be stretched in all directions and will still return to its former size and shape. Living systems are resilient if they can respond to shocks while maintaining core functions and relationships. Such systems reorganize and remain functional in the face of disturbance. Many ecosystems are periodically disrupted by storms or fires, for instance, but regenerate within weeks, months, or years to recover their former levels of biodiversity and productivity.

Disturbance is not an enemy to be avoided or combated within living systems. Rather, it is a partner in the dance of sustainability. Indeed, many ecosystems require severely disruptive events to maintain their long-term equilibrium. River deltas and other wetland ecosystems, for instance, may require periodic flooding to maintain their sediment, salinity, and nutrient levels. The flooding can be quite harmful to the current inhabitants of the deltas and wetlands. But the long-term flourishing of these ecosystems depends upon it. Likewise, lodgepole pine forests require intense fires to regenerate. The seed-bearing cones of lodgepole pines are sealed shut by resin. They can remain on standing trees for years, unopened and unproductive. For the seeds to be released, temperatures approaching 140 °F (60 °C) must melt the resin and open the cone scales. Only severe forest fires can generate temperatures like this within the crowns of standing trees.

Just as the integrity of democratic systems of government requires the periodic disturbance of elections to change out officials, so many ecological systems require disruptive events to maintain their core functions and cycles. This parallel between democratic and ecological systems underlines a key point. Resilience is less about holding onto a known and stable past than claiming a vibrant future. Certainly, the resilience of democratic governments comes from their capacity for creative disruption. The same could be said about ecosystems, economies, and human societies. Like ecosystems and economies, societies are networks of relationships. And like ecosystems and economies, societies prove their resilience when they undergo shocks without disintegrating. Resilient societies adapt to a changing world while maintaining core values and relationships.

William Ruckelshaus served on the Brundtland Commission. He was the first head of the US Environmental Protection Agency (EPA) in 1970 during the Nixon administration, and headed the EPA again under President Reagan. "Can we move nations and people in the direction of sustainability?" Ruckelshaus asked in 1989. He answered his own question with a keen sense of the stakes involved:

> Such a move would be a modification of society comparable in scale to only two other changes: the Agricultural Revolution of the late Neolithic and the Industrial Revolution of the past two centuries. Those revolutions were gradual, spontaneous, and largely unconscious. This one will have to be a fully conscious operation, guided by the best foresight that science can provide....If we actually do it, the undertaking will be absolutely unique in humanity's stay on the Earth.[31]

Ruckelshaus's poignant question places the task of sustainability – managing the scale and speed of change – squarely in deep time. It also underlines, to borrow Martin Luther King, Jr's evocative term, the "fierce urgency of now."

This chapter examined the history of sustainability for three reasons. First, the expanded time horizon of sustainability is Janus-faced. A heightened sense of responsibility for the future is complemented by a greater appreciation of the past. Janus' face was one of the most common images on Roman coins, perhaps because he was the god of choices. The implication is that the best choices are made when we, like Janus, look both to the past and to the future.

Second, creating sustainable societies is not simply a technological challenge. It is a cultural achievement grounded in the development of individual and social skills and virtues. These skills and virtues have deep historical roots. If we are to cultivate them anew today, we might well ground our efforts in the wisdom of ancient societies.

Third, the collapse of earlier societies helps us understand the dire consequences of business as usual. Some of the challenges that we face today are unprecedented. But many of our most pressing problems – such as resource depletion, ecological degradation, and overpopulation – have threatened human communities for thousands of years, and have destroyed many of them. If we fail to learn the lessons of history, we will be condemned to repeat its mistakes. Living in ignorance of yesterday puts tomorrow at risk.

Sandwiched between forerunners and future generations, we are temporary stewards of our lands, waters, and skies. Nature's bounty is not chattel to be used up, despoiled, or disposed of as we see fit. We inherit the earth as a trust. This inheritance comes with responsibility, an obligation to pass the planet on to future generations – minimally – in no worse shape than we received it.

If we succeed in creating sustainable societies, we will be averting a crisis and saving our collective skins. But to practice sustainability is not simply to pursue self-interest. It is a moral obligation. Senator Gaylord Nelson, who organized the first Earth Day in 1970, observed that "The ultimate test of man's conscience may be his willingness to sacrifice something today for future generations whose words of thanks will not be heard." Becoming a great ancestor is an ethical responsibility, and, one might say, a sacred trust. In the face of our current crisis, it is also a practical necessity.

Inquire and Explore

1. What is the world population today, and how high will it rise in your lifetime?
2. What are the limits to growth, and why do societies collapse?
3. What are your responsibilities to future generations, and how are they affected by climate change?

******

- Discover deep time and gain a future focus.
- Introduce yourself to organizations dedicated to fostering resilient human communities and ecosystems.

For web resources and more information on these topics, please visit conservationandcreativity.net

2
The Geography of Sustainability

To live sustainably is to conserve options for future generations while also appreciating one's biological and cultural
inheritance. To put the matter concisely: "Sustainability is
equity over time."[1] But sustainability extends moral concern
not only across expanses of time, but also across geographic
space. To practice sustainability is to move beyond the
national, political, economic, ideological, racial, ethnic, and
gender borders and cleavages that fragment and divide us.

The Brundtland Report stipulated that "concern for social
equity between generations...must logically be extended to
equity within generations."[2] In other words, reason dictates
that distant neighbors who share the world with us today
deserve our consideration no less than future generations who
will inherit the world. After all, our descendants may not live
in the same communities or nations as we do. Indeed, these
communities and nations may not exist in the distant future.
In turn, distant progeny will share very few of our genes, as
these are diminished by half with each subsequent generation.
Given such an exponential decline, within a few hundred
years our direct descendants will bear little more genetic
similarity to us than do strangers living on the other side of
the globe. So it is difficult to justify our concern for the
welfare of future generations based on patriotic fellowship or
genetic similarity. If we are willing to extend moral concern
in time beyond the near and dear, then we should also be
willing to extend moral concern in space.

From a sustainability perspective, the geographic extension of moral concern is a practical necessity. To effectively safeguard the welfare of descendants, we have to secure the welfare of contemporaries, whatever side of whichever border they currently occupy. "Protecting the rights of the most vulnerable members of our society," Aron Sachs observes, "is perhaps the best way we have of protecting the right of future generations to inherit a planet that is still worth inhabiting."[3] There are good moral reasons to protect the most vulnerable members of society from environmental degradation and injustice, apart from our sense of responsibility to future generations. But in an interdependent world, such protective action constitutes the only effective way to fulfill obligations to progeny. Today it is impossible to secure the welfare of future generations – and very difficult to secure our own welfare – by tending solely to our own backyards.

We begin this chapter by examining the notion of a global commons, the idea that the earth as a whole must become the object of caretaking. When we speak of globalization today, however, it is often not planetary protection that is the foremost concern, but rather the steady expansion of social, technological, political, economic, and cultural connections. These global interdependencies have both positive and negative repercussions. With this in mind, the chapter assesses the relationship of globalization to poverty, inequality, and social injustice, as well as opportunities to ground global caretaking in local, regional, and national relationships and initiatives.

The global commons

The Brundtland Report was entitled *Our Common Future*. The evocative title underlines our ever-increasing global interdependence. In earlier times, tribes, villages, and nations often existed in relative isolation. Today, global markets, media, travel, cultural exchange, technological linkages, and environmental concerns confirm our entwined fates. The future of humanity, for the first time in history, is truly a common one.

Pollution does not stop at national borders and climate change cannot be isolated to particular countries or continents. The same might be said about technology, media, and markets. Our socio-economic and ecological interdependencies bring both worldwide benefits and considerable dangers. Regardless of whether we enjoy these benefits, or suffer the problems, one thing is clear. The world is shrinking. And it is shrinking at a quickening pace.

Climate change dramatically illustrates our shrinking world and common future. Greenhouse gases and water vapor in the atmosphere trap heat on the planet in the same way that heat is trapped inside a greenhouse by its transparent ceiling. Carbon dioxide (CO_2) is the most important greenhouse gas. While other gases such as methane have many times the heat-trapping capacity of CO_2, much smaller amounts of these gases find their way into the atmosphere. Human activity around the globe puts about 37 billion tons (gigatons) of carbon dioxide into the atmosphere each year, primarily through the burning of fossil fuels and the destruction of forests, which otherwise would trap CO_2 in soils and trees.

The CO_2 in the atmosphere prior to the industrial age was about 280 parts per million (ppm), and it has been at that or a lower level for the last 800,000 years. James Hansen, perhaps the world's leading climatologist, announced in 2007 that 350 ppm of CO_2 was the maximum safe level. We are currently well over 400 ppm, and that number is rising steadily. Importantly, there is a greater difference in atmospheric carbon dioxide between today's levels and those of pre-industrial times than there was between pre-industrial times and levels during the last Ice Age. Clearly, we are in uncharted territory.

According to scientists at NASA's Goddard Institute for Space Studies, the average global temperature has increased by about 0.8 °C (1.4 °F) since 1880. Two thirds of this warming has occurred since 1975. The decade ending in 2010 was the hottest on record, with the preceding decade being the second hottest. Our current decade can already lay claim to the four hottest years on record, and that number is very likely to grow. While there will be much regional variation, global average temperatures may rise as much as 5 °C (almost 9 °F) by the end of this century. Such estimates are uncertain, but scientists concur that even a 2–3 °C rise in global temperature would have severe consequences. We are well on the way to catastrophe.

Regardless of the sources of greenhouse gases, climate change will impact all countries and continents. These impacts will be varied, and severe. Global warming threatens biodiversity, as countless species – perhaps a third of those that presently occupy the planet – will become extinct, unable to adapt to quickly changing climates. Over the coming decades, glaciers across the world will melt. Rivers fed by these glaciers which for thousands of years have provided water throughout summer seasons to irrigate crops and supply households will dry out. Thirst and hunger will be common, and disease will likely spread as clean, fresh water becomes scarcer. Rising temperatures will also decrease agricultural output in areas that currently benefit from optimal growing temperatures. In turn, global warming will cause ice in polar regions to melt, which, along with glacial melt and the thermal expansion of the oceans (as warmer water expands), will lead to rising sea levels. By the end of the century, sea levels may rise by 1–2 meters (3–6 feet). If all the ice covering Greenland and parts of Antarctica melts, the oceans could rise by 4–6 meters (up to 20 feet). Even conservative scientific estimates of sea-level rise predict the swamping of hundreds of islands and coastline communities. In turn, global warming will likely increase the intensity of meteorological disturbances, such as hurricanes, which cause flooding and storm damage, as well as other forms of extreme weather, such as severe, prolonged drought.[4]

It is probable that tens of millions of people – if not hundreds of millions – will become environmental refugees as a result of extreme weather, water and food shortages, ecological degradation, and coastal flooding caused by climate change. These refugees, desperately in need of water, food, housing, and jobs, will pose a daunting challenge to their homelands, to neighboring states, and to the international community. Social instability, civil strife, and violent, interstate conflict could significantly increase. National borders will not protect us from these dangers.[5]

The harms caused by climate change will not respect national sovereignty. And no country can solve the climate crisis on its own. Reducing fossil fuel use and forest destruction requires global cooperation. In turn, bringing atmospheric carbon dioxide down to a safe level entails sequestering hundreds of billions of tons of atmospheric carbon in plants, trees, and enriched soil.[6] The vastly increased acreages of grasslands and forests needed for this endeavor surpass the land available to any single nation. It will have to be distributed across the globe.

Climate change is perhaps the most prominent problem that is inherently global in its causes, its effects, and its potential solutions. But it is by no means the only reason that our world is shrinking. Depending upon your family history, the great-grandparents of your grandparents – people born in the mid- to late 1800s – grew up before the internet, television, radio, or telephone service could connect them to distant people. They lived prior to automobiles and airplanes that could transport them to distant towns or cities. They lived in a different world.

The great-grandparents of your grandparents lived prior to globalization. To be sure, they owned goods from other lands that were transported by merchant seamen who plied the oceans. In turn, colonialism had for centuries brought the affairs of distant peoples under the governmental control of imperial states. And in their efforts to gain such control, these states had, for centuries, fought wars in far-off lands. So the great-grandparents of your grandparents did experience global connections, and many were immigrants.

But there had never been a world war. There was no League of Nations or United Nations to provide a global forum for national governments. International non-governmental organizations – of which there are tens of thousands today – were nowhere to be found. Trade and other financial interactions, as well as cultural and technological exchanges between nations and peoples, were but small fractions of their current levels. The great-grandparents of your grandparents lived on the same planet as you do, but they lived before the earth had truly become one world.

In 1968, the *Apollo 8* astronauts – the first humans to have left the earth's orbit – took a picture of our planetary home as their command module rounded the moon. Their *Earthrise* photograph depicts a cloud-speckled planet, isolated upon the dark blanket of space, rising above a desolate moonscape. Dubbed "the most important photograph of the twentieth century," and possibly the most reproduced photograph of all time, this snapshot of a haven of life against the backdrop of a dark cosmos presented humanity with the first image from deep space of its common and unique home. No national borders were evident, just a singular, surprisingly small, and seemingly fragile planet.

The astronomer Fred Hoyle had earlier predicted that "Once a photograph of the Earth, taken from the outside is available...a new idea as powerful as any other in history will be let loose."[7] The *Earthrise* photograph of 1968 was indeed powerful. It helped people develop a personal relationship to the planet. It made patent the fact that we have a common home, a closed system with finite boundaries. The universe depicted in this photo looked like a pretty forbidding place, so, if the small haven of life we call earth were to survive and prosper, global caretaking would be needed.

Earlier that same year, Garrett Hardin posed a formidable challenge to the idea that collective caretaking of the planet was possible. Hardin, a biologist, published a now famous essay entitled "The Tragedy of the Commons" in the journal *Science*. The earth is indeed a closed system with finite resources, Hardin acknowledged. But these resources, precisely because they are viewed as a common inheritance, are likely to be exploited to the point of depletion, with catastrophic results.

Hardin tells the story of the English commons, public pastures where herders grazed their livestock. Each herder, wanting to make maximum use of the free forage, had an incentive to place more and more of his animals on the common land. Since all of the herders shared this same motivation, the commons very soon became jam-packed with livestock. Predictably, the land became severely eroded, with the once-luscious grass gnawed down to its roots. At that point, no new grass would grow on the bare, trampled soil. The commons became a hardscrabble wasteland.

In depicting the tragedy of the commons, Hardin was updating an insight first voiced by Aristotle thousands of years earlier. "That which is common to the greatest number has the least care bestowed upon it," Aristotle observed in *The Politics*. "Everyone thinks chiefly of his own, hardly at all of the common interest."[8] When he wrote these somber words, Aristotle was thinking about the common lands, buildings, and public spaces of the city of Athens in his native Greece. His point is even more relevant today, with a global commons shared by billions of people.

In 1968, when Hardin first published his essay, the global depletion of natural resources and the degradation of land, water, and air through overuse and pollution was a growing concern. Overpopulation was also much in the news. "Too many people...Too little food...A dying planet," was the succinct, dire conclusion of Paul Ehrlich's *The Population Bomb*, published that same year.[9] These concerns became widespread with the arrival of what was called "planetary consciousness" and "environmental awareness" in the late 1960s and early 1970s. Unfortunately, this shift in consciousness and growth of awareness did not reverse humanity's perilous course. Stocks of key resources, including biodiversity, continued to shrink rapidly over the coming decades. Many forms of environmental degradation and pollution grew worse. Meanwhile, the human population would double in size.

To avert tragedy, Hardin suggested that a sovereign power (that is, a world government) might be necessary to protect the global commons. Alternatively, the commons might be divided up into private landholdings where individual self-interest would ensure responsible, foresighted behavior. However, a brief survey of history demonstrates that neither sovereign states, with all their coercive powers, nor private landowners, with all their self-interest, have been consistently successful in protecting and preserving natural resources. But the news is not all bad. Both history and recent social science demonstrate that not all commons become overexploited. Indeed, even the English commons that Hardin referenced in his original article were often sustainably managed.

In 2009, American political scientist Elinor Ostrom received the Nobel Prize in Economics for investigating how common-pool resources – such as forests, fisheries, water supplies, and pastureland – can be well managed by large groups without a coercive central government and without being subdivided and appropriated as private property.[10] Voluntary associations of well-organized stakeholders are often quite successful at managing commons. To do so, they must establish clear rules by collective, participatory mechanisms. These rules must be adapted to local situations and be appropriately scaled, regularly monitored, and effectively enforced. The commons can be well managed, if its stakeholders are informed and dedicated. In this respect, the commons should be understood as a form of cooperative action, a verb rather than a noun.[11]

Ostrom does not suggest that there are no important roles for states and governments to play in protecting the environment, or that private ownership of resources is inherently unsustainable. But her pathbreaking scholarship did demonstrate that sustainably managing commons entails the development of a mixture of resource-management mechanisms, including those created and maintained by stakeholders who neither privately own, nor coercively control through a central government, the resources in question. This applies to individual stakeholders sharing common-pool resources such as pastureland, forests, rivers, and lakes. It also applies to nation-states, which have the collective responsibility to manage global resources, such as the oceans and the atmosphere. And it applies to other commons that are not primarily made up of natural resources, such as Wikipedia, Open Education Resources (OER), Creative Commons, and other internet sites and goods that are developed and managed collectively in the absence of a central authority.

Globalization and poverty

About the time the great-grandparents of your grandparents were born, Karl Marx and Friedrich Engels foresaw the rise of globalization. Observing the effects of coal-fired factories churning out marketable merchandise, cargo-laden steam ships plying the seven seas, railways and telegraphs crisscrossing nations whose forests were being felled and lands plowed under to produce more, and more varied, goods for consumption, Marx and Engels wrote:

> In place of the old wants, satisfied by the production of the country, we find new wants, requiring for their satisfaction the products of distant lands and climes. In place of the old

local and national seclusion and self-sufficiency, we have intercourse in every direction, universal inter-dependence of nations. And as in material, so also in intellectual production. The intellectual creations of individual nations become common property.[12]

To say that we now live in one world is to acknowledge – as Marx and Engels prophetically observed – that our interdependence has been steadily growing for the better part of two centuries.

The increased trade produced by European colonialism and imperialism in the eighteenth and nineteenth centuries initiated the modern development of economic globalization. The rise of American power after World War II intensified the process, as did the formation of transnational organizations mandated with securing and developing global finance and trade, such as the International Monetary Fund (IMF) and the International Bank for Reconstruction and Development formed in 1945 (which subsequently became part of a more expansive World Bank), and the General Agreement on Tariffs and Trade (GATT) forged in 1947 and superseded by the World Trade Organization (WTO) in 1995. The emergence of Russia and former Eastern bloc countries from behind the Iron Curtain in the 1990s, and the rise of China as an economic superpower in the 2000s, demonstrated that the last hurdles to a truly global marketplace had given way. Apart from a few isolated autocracies such as North Korea, the world's nations are now fully enmeshed in a global web of commerce.

Economic interdependence has increased markedly in recent years between nations, between economic sectors, between corporations, and between individual producers and consumers. Today, customers, workloads, and supply chains of materials are spread out across multiple continents, with the transnational buying and selling of goods accounting for an ever larger share of nations' commercial activity. Economically speaking, we have never been more connected.[13]

We can measure economic globalization by comparing the total dollar amount of a country's exports and imports to its Gross Domestic Product (GDP), the dollar worth of the domestic exchange of all goods and services. Worldwide, exports and imports are about two fifths as large as GDPs. This means that every time 5 dollars gets exchanged between the residents of a country, 2 dollars effectively crosses its borders. And in some countries, such as Germany, the ratio is approaching 1:1, with exports and imports nearly matching in value the goods and services exchanged within national borders.[14]

Globalization has its benefits. Heightened world trade has stimulated industrial efficiency and productivity. Global wealth, as measured by aggregated GDPs, has also increased. In turn, economic globalization has done a great deal to stimulate international cooperation, including international efforts to prevent conflict and war. As Thomas Friedman observed, "people embedded in major global supply chains don't want to fight old-time wars anymore. They want to make just-in-time deliveries of goods and services – and enjoy the rising standards of living that come with that."[15] But the heightened cooperation that economic globalization stimulates is not solely a product of people appreciating the benefits of peaceful barter over disruptive conflict. As globalization transforms strangers living on the other side of the planet into customers, clients, or suppliers of goods and services, concern for their safety and welfare may also develop. The local marketplace is not only a place of business; it presents an opportunity for human connection and care. The global market also bears this promise. Economic globalization stimulates increased cooperation to alleviate the human suffering caused by conflict, environmental degradation, health problems, and poverty.

But there are also downsides. The increased cooperation that globalization stimulates typically comes in response to problems that it has produced or worsened. International efforts to preserve biodiversity, for instance, face the vast destruction of habitat that occurs when countries grow crops for export on formerly forested land, and the pernicious effects of invasive species introduced through global transport and trade. Likewise, international organizations involved in efforts to combat climate change are grappling with a problem much exacerbated by economic globalization. The manufacturing, transport, and consumption of goods is the engine that drives economic globalization. Keeping that engine running depends on vast amounts of fossil fuels. Economic globalization is an accelerant of climate change.

A globalized economy is, by definition, an interconnected economy. Connectedness is often a good thing. But it has a seamy underside. Consider the global recession of 2008–10, a worldwide downturn in economic output that destroyed countless businesses, sparked economic turmoil in scores of nations, and caused massive job losses and unemployment. It began when bad lending practices contributed to an over-heated housing market in the United States. Soon enough the housing "bubble" burst, threatening a number of major financial institutions with insolvency. A panic on Wall Street quickly infected other financial markets. The dominoes started to fall, and before long US$50 trillion in global wealth had been wiped out. Despite the infusion of hundreds of billions of dollars in governmental bailouts, a global recession could not be averted.

Poor financial regulation in one nation (a practice unfortunately copied by other countries) created a grossly inflated real-estate market. When bad loans could not be repaid, the financial system imploded and economies collapsed. Tens of millions of people were thrown out of work, out of their homes, and into dire straits across the globe. Interconnected economies, if poorly managed, leave us prey to chain reactions that can quickly spiral out of control.

Globalization has other untoward consequences, particularly for those at the short end of the economic stick. The growing wealth produced through globalization was predicted to make everyone better off. The reality is that very little of this wealth actually trickles down to the impoverished masses, at least in comparison to the amount that is drawn upward, like iron filings to a magnet, to the fortunate few. As a result, vast economic inequalities have persisted, and in many cases have worsened. Five decades ago, the ten wealthiest countries in the world were 30 times as well off as the ten poorest. After 50 years of uninterrupted globalization, this disparity has doubled: the wealthiest states are now 60 times as rich as the poorest.

At the national level, wealth has also become very unevenly distributed and social inequalities have grown. The United States has been the foremost promoter of globalization. It remains one of the world's largest exporters, and is the largest importer.[16] At the same time, the United States has the most unequal distribution of wealth of any industrialized nation.

During the last two decades, top American CEOs have gone from earning about a hundred times more than the average full-time worker to earning more than 400 times as much.[17] The poverty rate in the United States from 2010 through 2014 has hovered at around 15 percent. One has to go back to the early 1960s to find a poverty rate significantly higher. And the number of Americans living in poverty is at an all-time high, approaching 50 million. American households with cash income of less than 2 dollars a day, the definition of extreme poverty, more than doubled between 1996 and 2011.[18] Poverty at this level is disempowering. It disables participation not only in economic affairs, but in social and political life as well.

Economic globalization heightens the production of wealth, at least periodically. But the fact remains that after a half-century of near-continuous growth in global trade, extreme poverty has not been eradicated. Not even close. We have – for the first time in history – reached the sad milestone of having one billion of the world's people unable to meet their most basic needs of food, fresh water, shelter, sanitation, and health care. That is because the wealth generated by global commerce is distributed so unevenly. It is estimated that for every US$100 in economic growth as measured by rising GDPs, only 0.6 percent, or 60 cents, finds its way into the hands of the poor.[19] As a consequence, the richest 1 percent of the world's people now owns over 40 percent of global assets, while the poorest half of the world's population owns barely 1 percent of global assets.[20]

In its *Global Trends 2015* report, the Central Intelligence Agency of the United States offered this assessment of the effects of globalization:

> The networked global economy will be driven by rapid and largely unrestricted flows of information, ideas, cultural values, capital, goods and services, and people: that is, globalization. This globalized economy will be a net contributor to increased political stability in the world in 2015, although its reach and benefits will not be universal. In contrast to the Industrial Revolution, the process of globalization is more

compressed. Its evolution will be rocky, marked by chronic financial volatility and a widening economic divide.[21]

In the game of globalization, the biggest losers, relatively speaking, are the world's poor. The biggest winners clearly are the large corporations engaging in finance and trade, and the individuals and nations that profit by them. Today, the wealth of the world's richest 500 individuals has surpassed the income of over half the world's population.[22] The distribution of the benefits of globalization has been staggeringly one-sided.

Dire poverty, as a rule, is not good for the environment. For one, it contributes to overpopulation. Owing to inadequate medical treatment, nutrition, and sanitation, the health and lives of impoverished children, particularly newborns, may be in chronic jeopardy. Consequently, impoverished couples have more children to increase the chances that enough offspring will survive. Moreover, in the absence of social security or other forms of state welfare, adult children of aging parents typically become primary caretakers, providing a safety net. Impoverished people need hedges against destitution in old age, and large numbers of children offer this security. Importantly, impoverished women generally lack the education and economic opportunities, and often the availability of contraception, which would allow them to exert effective control of their lives. Absent this control, they tend to become mothers of large families. As a consequence of these socio-economic forces, virtually all the population growth in the next century – well over 90 percent – is expected to occur in underdeveloped nations. And the poorest of the underdeveloped nations will experience some of the greatest increases in population.

Since the early 1970s, poverty has been identified as a "root cause" of overpopulation.[23] There are now four decades of data to support the claim that "development is the best contraceptive." Education and economic opportunity, particularly for women, along with sanitation, clean water and medicine to reduce child mortality, are the surest and fastest means to lower fertility rates and foster more demographically sustainable societies.[24] But neither education, economic opportunity, sanitation, clean water, or medicine is widely available in impoverished countries. In light of these facts, Vandana Shiva, an Indian environmentalist and feminist, argues that "giving people rights and access to resources so that they can regain their security and generate sustainable livelihoods is the only solution to environmental destruction and the population growth that accompanies it."[25]

Apart from overpopulation, dire poverty has other environmentally destructive effects. It can force people to hunt endangered species for food, or slash and burn rain forests to grow crops for subsistence or export. Slash-and-burn agriculture destroys critical habitat for wildlife and contributes significantly to climate change. Likewise, scavenging wood from already dwindling forests and eroded savannas for heating and cooking increases deforestation and desertification. But in the absence of more environmentally benign means of heating and cooking, the poor have little choice. "To care about the environment," renowned anthropologist and conservationist Richard Leakey observed, "requires at least one square meal a day."[26] Until basic needs are met, the demands of short-term survival typically overwhelm concern for environmental caretaking. Poverty and sustainability do not mix.

Social inequality and environmental justice

It may have been misleading to say that the great-grandparents of your grandparents lived before the world had truly become one world. Today's world, your world, is not really one world. Certainly, you live in a time of globalization. But the ills and benefits of globalization are very unequally distributed. Economic globalization has concentrated wealth in fewer and fewer hands. The effects of this inequity have been largely negative for the planet and its peoples.

Richard Wilkinson and Kate Pickett have conducted a thorough study of the social effects of the concentration of wealth. They observe that economic inequity has a very pernicious effect on societies, decreasing citizens' health, welfare, and happiness.[27] Developed countries with the lowest concentrations of wealth benefit from the improved mental and physical health of their citizens, higher life expectancy and lower infant mortality, higher levels of trust and social relations, higher educational performance, increased leisure time, and reduction in violence and imprisonment. Indeed, even in terms of social mobility – that is, the potential for individuals to climb up the income ladder to become wealthier than their parents – countries with lower levels of economic inequality generally fare better.

One might think of a high level of economic inequality as a form of pollution. Like other kinds of contamination that sully the land, water, and air, a high level of economic inequality degrades the quality of social life and is corrosive of many public goods, including political participation.[28] The data backing up this claim are very strong. In turn, studies have demonstrated that societies with greater social equality, including greater civil liberties and political rights, tend to benefit from improvements in environmental goods as well, such as access to clean water and sanitation, decreased levels of air and water pollution, and more stringent environmental protection policies.[29]

The Brundtland Commission concluded more than two decades ago that "inequality is the planet's main 'environmental' problem."[30] That is a sweeping statement. Socioeconomic life is a very complex phenomenon. While lower levels of social inequality clearly produce many social and environmental goods, other factors are in play. For instance, the trust citizens feel toward each other is an important factor. Levels of trust tend to be depressed in (racially and ethnically) diverse communities, even apart from the concentration of wealth.[31] So diverse societies may have a more difficult time producing safety, education, health care, environmental protection, and other public goods. This difficulty, however, further underscores the need to remedy the concentration of wealth. Social inequality also lowers levels of trust, and globalization increases social diversity. As communities and nations become ever more diverse, therefore, lessening economic inequality becomes an even more important means of fostering the trust that helps societies protect the environment and generate other public goods.

More empirical research is required before we can assert with certainty that inequality is the main environmental problem. But, as scholars investigating the available evidence observe, environmental degradation and social inequity are highly correlated. It is reasonable to conclude, in light of this evidence, that "social justice and environmental sustainability are inextricably linked, and that the achievement of the latter without greater commitment to the former will be exceptionally difficult."[32]

The concentration of wealth produces numerous social and environmental problems. But this does not mean that economic prosperity, or economic globalization, is wholly bad for individuals or the environment. Notwithstanding its contributions to inequality, economic globalization is positively correlated with higher life expectancy, probably because globalization increases access to health technologies, including water sanitation and medical treatment.[33] Rising income also corresponds to higher life expectancy and levels of happiness, though there are diminishing returns once income reaches a certain point.[34]

The diminishing returns of wealth occur because standard of living is only one of the ingredients of happiness. Much depends on the development of non-economic relationships – family, friends, and community. As importantly, people live comparatively. We typically assess our own welfare in relation to that of others. Our sense of satisfaction depends on our relative standing within a reference group. Indeed, people generally prefer to be absolutely poorer (or worse-looking) than they might be as long as they remain relatively richer (or better-looking) than most individuals in their reference group. In highly inegalitarian societies, such comparisons can undermine psychological health.[35] That may partially explain why citizens of the developed world have increased their wealth and level of consumption many times over the last 50 years, yet are no happier than their parents or grandparents were. Notwithstanding their greater wealth and consumption, somebody higher up the income ladder always owns and can spend more. Comparatively speaking, then, people are no better off, and they judge themselves accordingly. And if social inequality is increasing, with the rich getting richer and the poor getting poorer, an ever growing majority will perceive itself as worse off.

More egalitarian societies tend to do a better job protecting the environment. They also do a better job achieving environmental justice. Environmental justice refers to the equitable distribution of environmental hazards, such as pollution. It should come as no surprise that the wealthy and powerful of the world shield themselves to the greatest extent possible from environmental hazards that impact the quality of their lives. Why wouldn't they, given the resources at their disposal? It should also come as no surprise that the poor and disadvantaged of the world suffer a disproportionate share of existing environmental degradation, with the quality of their lives negatively affected.[36] As sociologist Ulrich Beck observed, while wealth accumulates at the top of the socioeconomic ladder, risks accumulate at the bottom.[37] The risks in question result from the inability to meet basic needs as well as the noxious impact of polluted environments.

In the early 1980s, environmental justice emerged as a potent issue in the United States. The disposal of toxic wastes in landfills or incinerators near low-income communities, which were predominantly African American in composition, sparked protests. A government report confirmed that hazardous waste landfills in a number of southeastern states were concentrated in regions populated primarily by African Americans.[38] In response to the charge of "environmental racism," the Environmental Protection Agency created an Office of Environmental Justice.

Environmental injustice is the unfair distribution of environmental benefits, costs, risks, and opportunities. It occurs whenever a particular group within society suffers disproportionate environmental degradation and danger, including the proximity of hazardous-waste dumps, or lack of access to various environmental goods, such as clean air and water. It also occurs whenever the natural resources of a community are extracted and sold without that community benefiting, or even having a say in the matter. Van Jones calls it "eco-apartheid...a situation in which you have ecological haves and have-nots."[39] Environmental injustice is very widespread, if not universal. As one study concluded: "In every nation of the world, poor people and minorities face greater environmental risks, have less access to environmental goods, and have less ability to control the environmental insults imposed on them."[40] While environmental injustice is suffered by underprivileged minorities and the poor within most if not all nations, it is particularly pronounced in highly inegalitarian societies. Environmental injustice also disproportionately affects the poorer nations.

For the better part of a century, the world was divided into rich, industrialized nations, the so-called "first world," and poor, undeveloped, and developing nations, once known as the "third world." As global consciousness grew in the 1960s, people learned that the fates of rich and poor nations were inextricably linked. "We travel together, passengers on a little spaceship," Adlai Stevenson famously said in 1965, "dependent upon its vulnerable reserve of air and soil; all committed for our safety to its security and peace; preserved from annihilation by the care, the work, and I will say, the love we give our fragile craft." To sustain Spaceship Earth, Stevenson argued, social inequities would have to be decreased: "We cannot maintain [the spaceship] half fortunate, half miserable, half confident, half despairing, half slave to the ancient enemies of mankind and half free in a liberation of resources undreamed of until this day. No craft, no crew, can travel safely with such vast contradictions. On their resolution depends the security of us all."[41] If social and environmental injustice remains rife on our planetary spaceship, Stevenson was saying, the craft is on a dead-end voyage.

When Stevenson gave his now-famous "Spaceship Earth" speech, he was acting as US ambassador to the United Nations. Almost four decades after his call for greater global equity in the service of planetary stewardship, the secretary-general of the United Nations, Kofi Annan, addressed the 2002 World Summit on Sustainable Development in Johannesburg, South Africa. Annan spoke of the "unsustainable practices [that] are woven deeply into the fabric of modern life." Those who suffer predominantly from these practices, Annan observed, may want to "rip up that fabric." The secretary-general proposed an alternative, non-violent remedy, one that entailed weaving "new strands of knowledge and cooperation." Nonetheless, Annan stipulated that "A path to prosperity that ravages the environment and leaves a majority of humankind behind in squalor will soon prove to be a dead-end road for everyone."[42] In an ecologically connected world, Annan agreed with Stevenson, great disparities in wealth and welfare will not only produce suffering for those at the bottom of the socio-economic ladder. Eventually, the ladder itself will break.

Annan's warning is timely. States that do not provide minimal security, the most basic needs, social justice, and political rights for their citizens are prone to revolution, as the upheavals across North Africa and the Middle East in 2011 demonstrated. The fact that these revolutions were made possible by the globalization of technology and culture, with social media such as Facebook and Twitter playing crucial roles in mobilizing mass protests, further underlines Stevenson's and Annan's observations about the impact of gross inequity in a connected world.

The more connected we are – socially, environmentally, and economically – the less tolerable will inequity be. Over two decades ago, the World Conservation Union put the point succinctly:

> Living sustainably must be a guiding principle for all the world's people, but it never will be while hundreds of millions live without enough of even the basic essentials of life. To make it possible for us all to think of the welfare of later generations and other species, we need a new kind of development that rapidly improves the quality of life for the disadvantaged. The Earth has its limits; with the best technology imaginable, they are not infinitely expandable. To live within those limits and see that those who now have least can soon get more, two things will need to be done: population growth must stop everywhere, and the rich must stabilize, and in some cases, reduce, their consumption of resources. Ways exist to do this without reducing the real quality of life.[43]

In the last two decades, the options have not changed. Unfortunately, neither has the growth in inequality and the unsustainable consumption of resources.

Sustaining glocality

The point of the previous chapter might be summarized by the maxim: learn from yesterday, prepare for tomorrow, act today. This complements the most popular slogan framing the issue of geographic sustainability: "Think globally, act locally." The imperative to act locally, like the imperative to act today, is at one level simply a matter of necessity: action can only occur in the here and now. While we may be focused on our responsibilities to the future, we can only take action in the present. Likewise, we can and should think of the planet as a whole. Still, as embodied creatures dependent on physical and emotional relationships that nurture and sustain us within local communities and ecosystems, we have to act in the geographically local.

Not all issues related to sustainability are global, or involve the global commons. Some problems are only resolvable at local, regional, or national levels. The potholes that make our streets hard to navigate demand local action, not planetary mandates. Likewise, many forms of pollution that issue from a single source and impact a restricted area demand local redress. Potholes and point-source pollution plague all countries. They are global problems in this sense. But that does not mean that they require, or can meaningfully be addressed by, a coordinated planetary response.

A good rule of thumb is that any problem that can be solved locally, should be solved locally.[44] We do not need to organize the Pothole League of the United Globe (aka PLUG) to improve our streets. Local governments and workforces can do the job much more efficiently. A global bureaucracy, sometimes derisively known as a globeaucracy, need not be created or get involved. Likewise, local, regional, or national governments are the best places to issue laws and regulations regarding point-source pollution that does not cross borders. Transferring problems to a broader social or geographic arena when it could be addressed locally generally decreases efficiency. More importantly, it weakens individual and community accountability. The more numerous and dispersed the stakeholders involved in a problem-solving task, the weaker the sense of individual and collective responsibility. Studies consistently demonstrate that a stranger in need is more frequently and consistently helped when he or she is encountered by an individual rather than a group of people, and by a small group more so than a large group. This phenomenon occurs, in part, because members of groups tend to pass the buck, assuming that someone else will take the time and make the effort to solve the problem at hand. This "bystander effect" is pervasive. A keen sense of responsibility – for our natural and social environment – is best cultivated by our direct participation in relatively small communities of stakeholders.

"Properly speaking," Wendell Berry observes, "global thinking is not possible. Those who have 'thought globally' (and among them the most successful have been imperial governments and multinational corporations) have done so by means of simplifications too extreme and oppressive to merit the name of thought."[45] Berry concludes that "The real work of planet-saving will be small, humble, and humbling, and (insofar as it involves love) pleasing and rewarding. Its jobs will be too many to count, too many to report, too many to be publicly noticed or rewarded, too small to make anyone rich or famous."[46] The moral weight of sustainability derives from our responsibility to sustain the community that sustains us. The communities that sustain us directly, and in a daily fashion, are primarily local. Without immersion in and devotion to the geographically near and dear – family and friends, colleagues and community, local agriculture, watersheds, and ecosystems – sustainability will remain an empty ideal.[47]

In the face of globalization, many parts of the world have been suffering a chronic "social recession," as ties to local community erode while social alienation grows.[48] This is an ominous development. To counteract this trend, advocates of sustainability often promote the localization of economies and cultures. A locavore movement, for example, promotes locally produced food. Home gardens are also encouraged, as are regional farms, farmers' markets, and Community Supported Agriculture (CSA) that provides local farmers a stable income for their produce.

Locavores often favor organic food production and the paying of living wages to agricultural workers. In turn, they prefer seasonal (rather than hothouse) fruits and vegetables. That is because local vegetables grown out of season in hothouses heated by fossil fuels contribute to climate change. A locavore strategy is typically meant to achieve four interrelated goals: the support of local economies and agriculture; healthy diets based on the consumption of fresh produce; the support of equitable economic relationships in the agricultural sector; and a small carbon footprint, as fresh food grown and sold locally does not require extensive fossil-fuel use for transportation, distribution, processing, and refrigeration.

The locavore movement is the agricultural face of bioregionalism. Bioregionalism is grounded on the principles that people should learn about and take responsibility for their local lands and waters, largely live within the means of local ecological systems, support local economies, and politically organize to promote the sustainability of their region.[49] Bioregionalism developed in the 1970s as part of the "back to the land" movement. In the face of growing consumer culture, urban congestion, suburban sprawl, and alienated mass society, people wanted to reclaim a sense of community and ecological connection to place. In some respects, bioregionalism is a reinvention of aboriginal lifestyles. Indigenous peoples have traditionally lived by bioregional principles and sensibilities. Their connection to the lands and waters that support their communities is economic, political, ethical, cultural, and spiritual.

Not all forms of localization are sustainable. Indeed, there is a name given to the pernicious localization of environmental concern: NIMBYism. NIMBY stands for Not In My Backyard. The term gained prominence in the 1980s in the wake of neighborhood efforts to oppose nearby sitings of waste dumps or incinerators. Neighborhood activists were not against landfills or incinerators per se. Nor were they trying to reduce, reuse, and recycle goods so that less waste would be generated and require disposal. They simply did not want the noxious waste sites to be located near their homes. Their solution was to transfer the problem to someone else's backyard.

NIMBYism safeguards the local by jeopardizing the geographically distant. It is a self-interested form of localism that shifts the burden of unsustainable lifestyles and economies. Oftentimes, backyards can be very large. NIMBYism frequently occurs at the national level. For many years, nations of the developed world shipped large portions of their hazardous wastes to less developed countries for disposal. In such situations, the poor and poorly represented of distant nations got the toxic end of the stick. Out of sight, out of mind.

The geography of sustainability brings what is out of sight back into mind. It is grounded in the conviction that there is no waste disposal port on Spaceship Earth. At a planetary level, you cannot throw something away because...there is no away.

The problem is that we live in a cradle-to-grave industrial system, also known as the take-make-waste world. Here, virgin natural resources are extracted and transformed into manufactured objects that ultimately end up in garbage dumps. This is a tremendous waste of resources, and the problem is worsened when waste that fails to find a grave comes back to haunt us. For instance, Americans throw away 100 billion plastic bags every year. That is the equivalent of 12 million barrels of oil. Only 2 percent of discarded bags get recycled. The rest goes to landfills or incinerators, or worse, fly about the countryside or end up in the oceans, where plastic entangles or is ingested by wildlife, killing a million birds, 100,000 marine mammals and turtles, and countless fish each year.[50]

As an alternative to the taking, making, wasting, and polluting of industrial societies, William McDonough and Michael Braungart envision a closed-loop, cradle-to-cradle system.[51] Here, sustainable product design and the use of natural and recyclable materials eliminate waste and pollution. At the end of their useful lives, durable goods are composted, becoming reintegrated into biological cycles, or after disassembly their parts are reused or recycled. In such a system, waste is no longer wasted.

Cradle-to-cradle systems reinforce the idea that there is no away. NIMBY gets translated into NIABY (Not In Anyone's Back Yard) or NOPE (Not On Planet Earth). Waste and pollution are not the only relevant issues in the building of sustainable societies. Various forms of industrial, commercial, agricultural, and residential development can also cause considerable harm to the health of the environment, to ecosystems, and to local populations. The challenge is to both heighten concern for our own stretch of the river while refusing to shunt risks and dangers to those living downstream.

At the same time, it is important that we don't go BANANAs (Build Absolutely Nothing Anywhere Near Anything). Reducing, reusing, and recycling are features of sustainable lifestyles.

But sustainability also entails building strong communities and economies. And as far back as Aristotle, we have evidence that the best and most durable societies are those with a large middle class. Here, citizens of moderate yet sufficient property and wealth can create accountable government through widespread participation. For this reason, Aristotle held democratic regimes to be "safer and more permanent" than oligarchies which existed in societies with great concentrations of wealth. In highly inegalitarian societies, Aristotle observed, the "state soon comes to an end."[52]

Societies often collapse because they ignore the dictates of environmental sustainability, degrading their lands and depleting natural resources. But they can also collapse because they fail to provide adequate welfare and maintain social cohesion. The French Revolution of 1789 demonstrated that the social system of the *ancien régime* – grounded as it was in the untrammeled power of the monarchy and nobility – was unsustainable. Likewise the Russian Revolution of 1917 proved its social system – grounded in the untrammeled power of the tzar and the unchecked power of a landed aristocracy – to be unsustainable. In both cases, large sections of society suffered greatly from poverty, neglect, and the abuse of power. When social revolutions topple governments and ruling classes, particularly if this occurs through violent means, there is no guarantee that the systems that emerge will be any more just or sustainable.

The CIA concluded its *Global Trends 2015* assessment with cautionary words regarding the growing gap that economic globalization creates between rich and poor peoples and nations: "Regions, countries, and groups feeling left behind will face deepening economic stagnation, political instability, and cultural alienation. They will foster political, ethnic, ideological, and religious extremism, along with the violence that often accompanies it."[53] With these concerns in mind, organizations such as the International Forum on Globalization (IFG), a research and educational institution, exposes, analyzes, and critiques the cultural, social, political, and environmental impacts of economic globalization.

The IFG also explores "alternative visions and policies to globalization that are more equitable, just, democratic, accountable, and sustainable for people and the planet."[54] These alternative visions and policies do not deny that the world is shrinking, or that the planet's peoples have a common future. Rather, they take on the task of fostering a widespread sense of responsibility for the global commons that stimulates rather than undermines a greater sense of connection, responsibility, and accountability to local communities and ecosystems. This is what is meant by glocality.

Glocality is based on the belief that making individual lifestyles and societies sustainable will require reinforcing connections between the local, regional, national, and global. The task is to build relationships and institutions that link global stewardship to concrete practices of local caretaking. Contemporary efforts to strengthen – and celebrate – bioregional relationships face a daunting challenge in a globalizing world. Glocality is a pathway to sustainability that makes the best of our connected world.

For millennia, the scale of social participation for human beings was limited to interactions within a small clan or tribe. Ecological relationships developed in a single watershed or landscape that was traversable in a few days of walking. Today, the communities of life that sustain us are greatly expanded. Socially, ecologically, and economically, they encircle the planet. Practically and emotionally, however, we are mostly sustained by relationships close at hand: family and friends, colleagues and fellow workers, customers, clients, sales and service people within our towns and cities, as well as relationships to local natural resources, species, and landscapes. Caring for these relationships is the stuff of sustainability.

Planetary consciousness is a tall order, but it is crucial if we are to maintain the health of the environment, and ensure sufficient social equity within and across nations to preserve core human values and public goods. At the same time, local attachments grounded in acts of personal caretaking are equally crucial. Sustainable societies cannot be achieved without individual investment in local community any more than future generations of humane citizens can be raised without parental love. The geography of sustainability is both personal and planetary. It is glocal.

Inquire and Explore

1. Why is climate change occurring, and what will its impacts be?
2. What are the costs and benefits of globalization, and how can you live more glocally?
3. Is the global challenge of ending poverty succeeding?

* * * * * *

- Discover the adverse impacts of high levels of economic inequality.
- Learn why "there is no away" in our interdependent world.

 For web resources and more information on these topics, please visit conservationandcreativity.net

3
Ecological Resilience and Environmental Health

Sustainability is grounded in the responsibility we have to sustain the community that sustains us. Most fundamentally, as biological organisms, we are sustained by the biosphere. Without the web of life to support us, we would quickly perish. Our livelihoods and lives are wholly dependent upon a nourishing natural environment and its resources – a stable climate, clean air to breathe, fresh water to drink, wash, and utilize for agriculture and industry, fertile soils in which to grow our food, grasslands for pasture, forests to provide our lumber and other goods, and lands and seas that supply us with plants and animals for food and countless other materials that we employ in our manufacturing and commerce. But nature does not only sustain us by way of resources we consume. The earth is blessed with a tremendous variety of landscapes and waterscapes, and a vast diversity of species, both plant and animal. This cornucopia of life sustains our recreational, emotional, artistic, and spiritual needs.

The economy, it is often said, is a wholly owned subsidiary of the environment. The same might be said of human civilization. Our highly urbanized existence, technological lifestyles, and complex cultural practices may mislead us into thinking that we have left the natural world behind. But natural resources and vibrant ecosystems still provide the crucial foundation for every human society and economy.

Today, this truth is all too evident as we confront the prospect of maintaining civilization on a warming planet whose natural resources are dwindling. Protecting the web of life – safeguarding ecosystems and healthy environments – is the *sine qua non* of sustainability.

In this chapter, we explore the ecological basis of sustainability. We start with the relationship between biodiversity and the resilience of ecological systems. Resilient ecosystems provide highly beneficial resources and services, and it is often useful to assess these benefits in economic terms. But nature is not only, or even primarily, of economic value. It is also something we cherish, and this love for nature is crucial to its safeguarding. We conclude by examining the nature of a healthy environment, understood as an unpolluted, biologically enriched, adaptive system.

Biodiversity and resilience

In the 1800s, the passenger pigeon was one of the most abundant species in the world. Vast flocks of these birds filled the skies of the United States for as far as the eye could see. Their nesting grounds often spanned hundreds of square miles and contained tens of millions of birds. But its gregarious nature and abundance made the passenger pigeon easy prey. At some nesting sites, as many as 50,000 birds were killed each day in organized hunts that lasted for months. While far-sighted individuals attempted to restrain the slaughter, their efforts proved too little and too late. Flocks upon flocks of the birds were hunted down, until there were no more to be found. The last passenger pigeon died in captivity at the Cincinnati Zoo in 1919.

The passenger pigeon is just one among many hundreds of species of plants and animals that have been brought to extinction by human hands. The eradication of a species is not always the product of excessive hunting or harvesting. The greatest threat to most species today is not direct predation, but the destruction of their natural habitats as forests and grasslands are developed for agricultural, industrial, or commercial purposes. Other prominent causes of the loss of biodiversity include the introduction of invasive species, pollution, and climate change.

The extinction of species is a natural phenomenon. Plants and animals regularly die out when they fail successfully to adapt and reproduce within their changing environments. This is natural selection at work. Scientists call it the background or natural extinction rate. Of course, new species are also always evolving, gaining footholds, and propagating, again as a product of natural selection. When the rate of extinction rises markedly above the rate of speciation during a set period of time, scientists call this a mass extinction or extinction event. The threshold for such an event – how large a drop in species there has to be to qualify as a mass extinction – is open to debate. But within the last 540 million years, fossil records indicate that at least five mass extinctions occurred.

In the Permian-Triassic extinction, sometimes known as the Great Dying, up to 96 percent of all marine species and over 70 percent of terrestrial vertebrates perished. Occurring about 250 million years ago, it may have been caused by climate change, a meteor or asteroid impact, or increased volcanic activity. The most recent mass extinction, some 65 million years ago, brought about the end of the dinosaurs, likely the result of a meteor impact. We are currently in the midst of a sixth extinction event. The difference between the current event and the preceding mass extinctions is that today's crisis is anthropogenic, the product of a single species: *Homo sapiens*.

The Millennium Ecosystem Assessment, carried out under the auspices of the United Nations Environment Program (UNEP), brought together more than 1,300 scientists from 95 nations to assess the extent, causes, and consequences of ecological degradation and species loss. Completed in 2005, the report noted that more forests and grasslands were converted to agriculture in the last 60 years than in the previous two centuries. In turn, the last half-century has seen unprecedented levels of other disturbances, including accelerating urban and suburban growth; the increased hunting, fishing, and harvesting of animals and plants; proliferating international trade and travel, which introduce invasive exotic species; pollution of lands, waters, and skies; and the burning of fossil fuels, which contributes to climate change. All of these factors have contributed to the extinction of species and ecological degradation.

It is difficult to predict how many species ultimately will perish as a result of overconsumption, habitat loss, pollution, climate change, and other human-caused impacts. Part of the problem is that we do not even know how many species currently exist. Estimates range from 3 million to 100 million, with recent studies suggesting a plausible figure to be between 8 and 10 million.[1]

The International Union for the Conservation of Nature (IUCN) currently lists 17,000 species as threatened with extinction. That includes a tenth of the world's birds, a fifth of its mammals, almost a third of all amphibians, and more than a third of freshwater fish.[2] By the middle of this century, most of the ocean's large predatory fish may also have disappeared. They will have become victims of our appetites. While much attention goes to fish that are prominent on our menus, such as cod, haddock, tuna, grouper, and swordfish, studies now show that even the small forage fish that larger fishes consume, such as sardines, herring, and menhaden, are in decline. The sharp drop in the world's populations of forage fish is largely the consequence of their use as fishmeal for aquaculture and animal feed.[3]

Scientists estimate that as many as half of all known mammal and bird species may be extinct within 300 years. More alarmingly, some studies suggest that half of all species may become extinct within the next century. Whatever the toll, clearly we are in the midst of a crisis. Somewhere across the globe, a species goes extinct every 20 minutes.[4] This extinction rate is estimated to be 100–1,000 times greater than the natural background rate.[5] Whether our current crisis will match or surpass the mass extinctions that have occurred in the last 500 million years is unknown, and rather beside the point. The devastation of biodiversity – a crime against the planet and future generations – is under way. No planetary extinction event has ever wiped out all forms of life. The current mass extinction, even if we do nothing to avert it, will be far from total. But there is no doubt that a planet deprived of its rich biodiversity – an earth inhabited by abundant viruses, bacteria, algae, cockroaches, and termites but few fish, reptiles, amphibians, birds, or mammals is a tragedy of massive proportions.

The tragedy is compounded because the current ecological crisis is not only a crisis of extinction. It is a crisis of speciation. The problem is not only that we are causing plants and animals to die out. The problem is that much of the planet is either under human occupation or suffering from human use and abuse. There is insufficient (healthy) habitat to facilitate the evolution of new species.

The biological history of our planet has been one of extinction and speciation. Indeed, the extinction of one (set of) species often created the ecological space needed for a new (set of) species. The development of many mammal species, for example, was much abetted by the disappearance of the dinosaurs at the end of the Cretaceous period. With the departure of these large reptiles 65 million years ago, habitat and resources became available for mammals and a dramatic rise in their speciation occurred.

Today, however, the extinction of plants and animals is not creating ecological space for new species because that space is occupied or cultivated by human beings. "For the first time in hundreds of millions of years," conservation biologist Michael Soulé concludes, "significant evolutionary change in most higher organisms is coming to a screeching halt."[6] By higher organisms, Soulé means vertebrates in general and mammals in particular. While a planet largely colonized by the human species will not much restrict the ongoing evolution of bacteria and viruses, algae, and some plants and insects, larger and more complex organisms require expansive habitats to procreate and flourish. Absent these expansive habitats, neither the preservation of large, complex species nor the evolution of new ones is likely.

One way to protect biodiversity and maintain habitat for wildlife is to establish parks and preserves. Today, about an eighth of the earth's land surface is designated as nature preserves or parks. Unfortunately, active protection and management on much of this land is often scant, so the poaching of wild animals and plants and the destruction of habitat continue. In turn, protective efforts that do not account for the needs of peoples whose livelihoods are traditionally derived from the land often prove ineffective and inhumane. As much as half of all lands designated for conservation over the last century was formerly occupied by indigenous peoples or regularly utilized by them. Many millions of these native and other land-based peoples have been displaced.[7] To be effective, conservation must balance the need for protected habitat with the fostering of sustainable livelihoods for indigenous and local peoples.

The active, well-managed protection of large areas of land (and sea) is crucial to the preservation of biodiversity. However, as Lester Brown observes, "The traditional approach to protecting biological diversity by building a fence around an area and calling it a park or nature preserve is no longer sufficient. If we cannot also stabilize population and climate, there is not an ecosystem on earth that we can save, no matter how high the fence."[8] In an era of climate change, high fences may actually prevent animals, and even some plants, from migrating to (cooler) altitudes or latitudes better able to sustain them.

A 1°C change in temperature is equivalent to a latitude change of 100–150 kilometers. Given the predicted rise in global temperatures over the next century, plants and animals may have to migrate several hundred kilometers toward the poles to stay within their maximum thermal tolerance levels. Many species will not be able to migrate toward the poles (or higher in altitude) fast enough to survive.[9] Oftentimes, cooler habitats will be occupied by humans or will be otherwise unavailable to migrating species. Protecting biodiversity in an age of climate change will require wildlife "corridors," extensive stretches of protected land largely running north–south that will allow the unimpeded migration of species facing thermal stress. It goes without saying that establishing extensive wildlife corridors in a world of growing human populations and land use will not be easy.

Whether we can establish and maintain sufficient habitat and corridors to preserve existing species and allow the evolution of new species is an open question. Currently, we are going in the wrong direction. Biodiversity is rapidly dwindling. E. O. Wilson, a biologist and conservationist, observed that "The one process now going on that will take millions of years to correct is the loss of genetic and species diversity by the destruction of natural habitats. This is the folly our descendants are least likely to forgive."[10] Wilson wrote these words in the early 1980s. Since then, we have learned that climate change may contribute even more than habitat loss to the extinction of species. Indeed, climate change alone may reduce biodiversity by a third.[11]

Why should we, or our descendants, care about biodiversity? What would it matter if there were a couple million fewer kinds of plants and animals on the planet? Would it really be so bad living in a world without sperm whales and sea otters, black rhinoceroses and long-snouted bats, great white sharks and red groupers, spotted pond turtles and tiger chameleons, nutmeg trees and redwoods, paradise parakeets and moustached kingfishers, blind cave beetles and water fleas, octopus coral and spiny-tail crayfish, banded horned tree frogs and spot-tailed warty newts? Most people might not even notice the difference.

Certainly, it would not be the end of the world if these endangered species disappeared. But it would be a much poorer world, and a less resilient one. A species within an ecosystem has been likened to a rivet holding together the wing of an airplane. You can pull out a few rivets and still have a functional airplane. But, at a critical point, the loss of one rivet too many will cause a wing to fall off, with catastrophic results for the airplane and its passengers.[12] The analogy suggests that resilient ecosystems have built-in redundancy. Some of their species can be lost without core functions and relationships being destroyed. Given time to recuperate, ecosystems will often rebound in inspiring ways after major shocks.[13] But there is a tipping point. And we do not know how many rivets have already been lost or whether popping the next one will prove catastrophic.

The analogy of species within an ecosystem to rivets on an airplane wing is somewhat misleading. First, not all species within an ecosystem are as crucial to its overall health as others. The loss of a single "keystone" or "driver" species may undo an entire ecosystem, just as the loss of a keystone (the wedge-shaped stone at the top of a vault or arch in a wall) can cause the collapse of an entire building. Second, the resilience of an ecosystem does not depend simply upon the number of species inhabiting it, but upon the number of strong interactions between these species. Resilience is gained through self-reinforcing connections among species. Unlike rivets on an airplane, species within ecosystems interact with one another, and evolve together. It is the complex nature of this interaction and evolutionary change that makes for the health and strength of ecosystems.

Still, you cannot have strong, self-reinforcing interactions between species if you do not have diverse species to do the interacting. So species diversity remains a reasonable means of assessing the resilience of an ecosystem. And the loss of a species almost always signals a decline in resilience.

The value of ecosystems and their services

The extinction of a species may sadden us deeply at a personal level. We may grieve for the loss of one of the members of the natural community in the same way we grieve for the loss of a member of the human community. But the extinction of a species is not only an emotional harm. It has an impact on the resilience of ecosystems and, by extension, the resilience of the planet's life support system. The loss of biodiversity threatens human welfare. Oftentimes, it has a direct impact on economic life.

Since ancient times, astute observers have tallied the economic costs of the destruction and degradation of nature. In the *Critias*, Plato surveyed the damage done to ancient Greece by leveled forests and eroded land:

> By comparison with the original territory, what is left now is, so to say, the skeleton of a body wasted by disease; the rich, soft soil has been carried off and only the bare framework of the district left....What we now call the plains of Phelleus were covered with rich soil, and there was abundant timber on the mountains....The soil got the benefit of the yearly "water from Zeus," which was not lost, as it is today, by running off a barren ground to the sea.[14]

In a similar vein, George Perkins Marsh assessed the damage done by two and a half millennia of civilization to Mediterranean lands, and observed that the same loss of nature's bounty through deforestation was occurring at a quickening pace in America. In his 1864 book, *Man and Nature*, Marsh noted the tremendous "service" done by trees, insects, and "minute organisms" that stabilize climate, retain freshwater, buffer winds, and decompose matter.[15]

A century later, scientists began the arduous task of quantifying the extent to which the human economy is quite literally a subsidiary of the natural environment. In 1981, Paul and Anne Ehrlich coined the term "ecosystem services" to denote the countless ways that the natural world provides benefit to humanity.[16] They were primarily concerned with the loss of biodiversity. However, ecosystem services may decline even in the absence of extinctions. The loss of certain wetlands, forests, or grasslands may not in themselves cause extinctions if the species found within them are plentiful elsewhere. Yet such losses may produce a dramatic decline in ecosystem services to their respective regions.

Ecosystems provide an extensive range of services. These services include the pollination of crops and natural vegetation, seed dispersal, erosion control, air and water filtering and purification, soil generation and soil fertility renewal, climate regulation and carbon sequestration, pest control, waste decomposition and detoxification, provision of pharmaceuticals, wild food and game habitat, flood control, pollution abatement, protection from ultraviolet solar radiation, moderation of temperatures and winds, recreational opportunities, and the provision of aesthetic beauty, intellectual stimulation, and spiritual renewal.

Consider the economic impact of losing just one ecosystem service: pollination. Bees pollinate many of the crops that we grow. As much as 30 percent of our food production depends on their efforts. Unfortunately, bee populations in many parts of the world are in decline, so we may be faced with the prospect of paying to achieve artificially the pollination that nature once provided free of charge. And, of course, bees do not only pollinate our crops. They also pollinate, and help to preserve, vast numbers of non-agricultural species of plants. These plants serve as anchors for lengthy food chains. So the decline of bee populations threatens both the bottom line for crop growers as well as the health of entire ecosystems. The value of bees, and other insect pollinators, goes well beyond the money they save farmers, orchardists, and ultimately consumers. Even so, their value as pollinators has been calculated at over US$200 billion annually.[17]

Beginning in the 1990s, efforts were made to estimate the economic benefits we accrue from the various services that nature provides. In a seminal paper published in the prestigious journal *Nature* in 1997, Robert Costanza, along with other ecologists and economists, assessed the monetary value of the world's ecosystem services. They arrived at a figure of US$33 trillion (and possibly as high as US$54 trillion) for 17 key ecosystem services at then current prices.[18] Importantly, this figure did not include the monetary value of all the non-renewable resources that nature provides, such as coal, oil, gas, and minerals.

The figure of US$33 trillion was 1.8 times the then global gross national product (the total dollar value of all goods and services bought and sold that year). In other words, if the services that nature provides free of charge on an ongoing basis were denied to us, we would have to double our economic output just to maintain our current level of welfare. Of course, basic survival – let alone maintaining or doubling economic output – would be a very tough challenge were nature's services to end.

Gretchen Daily's edited volume, *Nature's Services*, was published the same year as Costanza's work. Daily and a large group of collaborators tackled the thorny issue of the value we should place on – and the price we should be willing to pay to protect – the "natural life-support system and processes" that sustain our existence.[19] Within a decade, hundreds of academic papers devoted to the examination and valuation of ecosystem services were being published each year. The methods employed to make Costanza's and Daily's calculations, and the final figures reached, have been subjected to much critical analysis. Whatever their methodological shortfalls, these and later studies challenge conservationists and economists to grapple with a stubborn fact: the monetary value of the goods and services that nature provides – and will cease to provide if human-caused losses to biodiversity and environmental destruction continues – is vast.

The Millennium Ecosystem Assessment revealed that up to two thirds of the world's ecosystem services are being significantly degraded. We cannot afford their loss. If we had to replace these natural services with artificial means we would be pushed far beyond our technological capacities. And, of course, many ecosystem services are virtually irreplaceable, regardless of how much economic muscle and technological ingenuity we might expend. The climate-regulating effects of the atmosphere provide one example of an irreplaceable service. Absent its benefits, civilization would not survive long enough to begin the task of investing money and ingenuity into the development of a viable substitute.

Efforts to analyze the economics of ecosystem services have improved markedly over the years. Still, monetary values remain difficult to estimate and margins of error are considerable.[20] In part, this is because economic analyses cannot accurately calculate the future value of ecosystem services. Preserving biodiversity, for instance, both benefits us in the here and now and conserves options for the future.[21] Valuable pharmaceuticals might be developed from now-threatened plants. We cannot well predict how many such discoveries will be made, or what their economic impacts will be.

Likewise, the economic value of any particular species cannot be reliably calculated simply by examining the services that it directly provides. That is because the removal of one strand from the web of life weakens the whole. Any number of interdependent ecosystem services may be degraded by the extinction of a single species, even if that species is not directly involved in the provision of a known service. The costs we incur today when we preserve a particular species or maintain an ecosystem's health may not have obvious monetary benefits. Still, it is likely a wise investment.

The investment in ecosystem services paid immediate dividends to New York City. Its water system is the largest in the nation (using 1.3 billion gallons a day), and one of a small number that is unfiltered. The Catskills/Delaware watershed long provided New Yorkers with this clean drinking water. However, sewage and pesticides eventually degraded the area's quality of the water below national standards. Officials considered building a filtration plant, which would require an initial investment of up to US$8 billion and cost hundreds of millions of dollars annually to operate.

Instead, New York City invested in the caretaking of the Catskills/Delaware watershed. It acquired and protected hundreds of thousands of acres of land, upgraded wastewater treatment plants, and extensively monitored the area.[22] The result was an ongoing supply of clean drinking water at a fraction of the cost of building and maintaining a filtration plant.

In a similar vein, some national governments, such as Costa Rica's, pay landowners to protect ecosystems on their property. At a transnational level, the United Nations organizes a collaborative program, UN-REDD, aimed at Reducing Emissions from Deforestation and Forest Degradation in Developing Countries. Forest degradation and destruction account for almost 20 percent of greenhouse gas emissions. That is more than is produced by the burning of gasoline, diesel, and jet fuel in all the world's cars, trucks, trains, and planes. REDD agencies calculate the financial value of the carbon that is stored in a developing nation's forests. They then work with developed countries to provide monetary incentives for the preservation and sustainable management of these threatened forests. Worldwide, the "trading of ecosystem services," such as carbon sequestration, is rapidly gaining ground. Stock exchanges are now heavily involved in the economics of ecosystem services.[23]

However we do the math, nature provides us tremendous economic benefits. As we learn to determine the economic value of ecosystem services more accurately, however, it is well to remember that everything of value cannot be priced. We tend to value goods and services according to their (fluctuating) cost in the marketplace. But some things are priceless, or at least are much devalued if not wholly corrupted by being reduced to dollars and cents.[24] Human rights fall into this category. Freedom of association and freedom of speech certainly have economic impacts. One could attempt to assess these costs and benefits, and perhaps even derive a monetary figure for maintaining these rights in any particular country. But we betray our understanding of the true value of human rights were we able to buy and sell them as commodities in the marketplace.

Likewise, biodiversity and a healthy environment ought not to be reduced to monetary values. That is because we have a moral duty to preserve species and a healthy environment, just as we have a moral duty to establish and maintain human rights, regardless of their economic impact. This does not mean that economic considerations are off the table. If we partake of economic life, then we participate, directly or indirectly, in the monetary valuation of things that we might otherwise be loath to reduce to a cash figure. After all, we could save countless lives (not to mention liberties) in our own countries and across the globe by investing more money in medical care, nutrition, clean air, and clean water. We could save tens of thousands of lives and prevent untold injuries simply by mandating lower speed limits (which significantly reduce traffic fatalities) or insisting on higher safety standards in workplaces. But we often choose not to do these things. Though we may be reluctant to admit it, we often trade human health, welfare, and life for the sake of efficiency or financial gain. For better or worse, economics plays a large part in the choices we make. That is why the practice of sustainability maintains economics as one of its central concerns. The point, however, is that economic considerations must remain in balance with the safeguarding of other goods, chief among which is ecological health.

Biocentrism and biophilia

Calculating the economic value of nature, however fraught with difficulties and dangers, is increasingly important to the effort of building sustainable societies. Yet it reinforces the tendency to evaluate everything in terms of its costs and benefits to a single species: our own. Duties to humans of this and future generations are important, no doubt. But do they exhaust the breadth of our responsibilities? To think always and only from the point of view of human needs, wants, and benefits is to be anthropocentrically oriented (*anthropos* is the Greek word for human being). Biocentrists maintain that our purview should be expanded. They insist that we protect and preserve nature not only or primarily because it serves human needs and wants, but for its own sake.

An egocentrist is someone concerned solely with his or her own welfare. An egocentrist might occasionally be kind to another person if doing so also serves his or her interests. But such an action would only be a means to a self-serving end. Moral systems push us beyond egocentrism. They maintain that we have duties to other people, and should act with their good in mind. We have an obligation to save a toddler drowning in a pool, for instance, even if doing so would ruin our good clothes or cause us to miss our next appointment. In the same vein, biocentrists argue, we have a duty to protect and preserve nature even when it does not serve our economic interests or provide us with any instrumental value.

There is a vast philosophic literature within the field of environmental ethics that addresses the issue of moral duties to the natural world. The issues raised go beyond idle speculation. They have, in most countries, already impacted public policy. Animal rights are a case in point. Laws forbidding cruelty to animals are grounded on a presumed moral obligation to other species. The field of environmental ethics explores, among other things, how far our duties to the natural world extend, and how these duties differ from those we have to fellow human beings. Here, we only touch upon these issues in order to tie the intrinsic value of nature to the pursuit of sustainability.

Bill Mollison argues that "Sustainable societies emphasize the duties and responsibilities of people to nature equal to those of people to people."[25] Mollison's assertion would probably not be endorsed by the majority of those who consider themselves practitioners of sustainability. Still, it forces us to ask a key question: can sustainable societies be built in the absence of any sense of moral duty to the natural world? It may be mistaken to describe our relationship to the natural world with the same ethical concepts and categories we employ to describe our relationship to the human world. Indeed, it may be a form of anthropomorphism – the transposing of human characteristics and values onto non-human beings – to saddle nature with our ethical categories in the first place. Certainly, it presents us with some challenging quandaries.

If our species has moral obligations to nature, we might ask, do similar obligations exist among other species? Infanticide and cannibalism are ethically abhorrent acts that should be forbidden. But such practices are carried out by many non-human species. Should we intercede, then, to prevent fish, spiders, snakes, rodents, lions, bears, and apes from engaging in cannibalism and infanticide? And if we believe ourselves to be morally obligated to jump in a pool to save a drowning child, do we also have a moral obligation to save chipmunks from forest fires, drowning earthworms from drenching storms, and antelopes from hungry lions? If these prospects seem absurd – or at minimum a slippery slope to insoluble ethical conundrums – then we have to acknowledge that the language of ethics does not seamlessly describe our relationship to the natural world. We cannot simply extend the specific terms and conditions of our moral duties and legal responsibilities to encompass all the non-human participants in the web of life.

We need not impose our ethical framework on the natural world in order to respect it. Indeed, the deepest form of respect for nature may arise when we treat it not as an extension of our own moral universe, but as distinct. Perhaps nature deserves our respect precisely because it escapes the ethical categories that we have developed to order and honor the human world.

Morality can be misused. Egregious cases of domination, brutality, and genocide have resulted from one people imposing its ethical commitments (often grounded in religious beliefs) upon another. And even when greed or the lust for power is the root of such inhumanities, moral justifications are often developed to translate might into right. So we have to be very careful whenever we employ our moral commitments to order the world. This is especially true regarding other species.

At the same time, a deep respect for nature – whether or not it translates into the belief that we have specific moral duties to other species, or that animals (and perhaps plants) have specific rights that approach or are equal to those we accord human beings – appears crucial to the practice of sustainability. To be sure, there is nothing illogical about a sustainable society that simply views the natural world as a stock of resources to be exploited and managed in a frugal manner. But we might well wonder whether such a society would actually develop in the absence of an abiding respect for nature and the accompanying aesthetic and emotional relationships. Sustainability demands an expanded temporal and geographical horizon. This expansion of concern in space and time requires seeing oneself as part of a greater whole. Nature is the most encompassing whole we can directly experience with our five senses.

The seventh-generation mandate of the Iroquois nation, for example, was grounded in a broad – one might say, spiritual – connection to nature. Every strand of the web of life was sacred. The Lakota Sioux employed an expression in their prayers and rituals, *Mitakue Oyasin*. It can be translated as "We are all related." The phrase was directed not only to fellow tribal members, but also to the plants and animals that shared the world of the Sioux. *Mitakue Oyasin* emphasized the underlying connection between all living beings, the oneness of life. Without these commitments to a sacred unity, native peoples' deep sense of responsibility for the welfare of future generations may not have developed.

E. O. Wilson argues that a love of life grounded in the fundamental relatedness to all things biological is central to our being. Wilson insists that the love of nature is in our DNA. He coined the term "biophilia" to capture this genetic predisposition of humans to affiliate with other forms of life, to love, and take pleasure in, the natural world.[26] Having evolved over millions of years in close proximity to other species and their ecosystems, and being wholly dependent upon them for survival and welfare, human beings came to appreciate and enjoy nature. We instinctively desire food, drink, and air to breathe. Likewise, Wilson insists, we also desire to be surrounded by and interact with natural things. Human beings are, quite naturally, lovers of nature. In aboriginal cultures, this love of nature was manifested in song and dance, ceremony and prayer. Homage would be paid to the animals whose lives were sacrificed to satisfy human needs. Nature was honored, revered, and loved – even, or rather especially, when it was consumed.

Biophilia faces an uphill climb in the modern world. Frequent and extended interactions with the natural world is the best way to foster the emotional attachments of biophilia. But ensuring frequent and extensive interactions with nature is no small feat in a world of seven billion people. On such a crowded planet, wilderness is quickly losing ground to urban growth, suburban sprawl, and agricultural development. And our urbanized and technological existence – filled as it is with manufactured goods, industrial processes, artificial products, electronic media, and virtual reality – increasingly divorces us from natural cycles, biological processes, and living things.

In the modern world, biophilia may well atrophy, in the same way that our natural predisposition for vigorous physical exercise and face-to-face community might atrophy in a world of automobiles, office cubicles, and homes with WiFi and entertainment centers. As Stephen Kellert observes, the desire to be in nature, though innate, constitutes a "weak genetic tendency whose full and functional development depends on sufficient experience, learning, and cultural support."[27] We did not evolve to live the life of reclusive couch potatoes or video gamers cut off from both nature and face-to-face human community. But our technology facilitates this cultural development or, should we say, cultural diminishment.

Ecological caretaking serves our long-term self-interest. But as the lives of addicts attest – and as we all know first hand, given some of the things that we eat, drink, smoke, say, do, or neglect to do – human beings often fail to act in their long-term self-interest. Sometimes we need help to ensure that we pursue our own good. In this respect, biophilia may not simply be a predisposition to surround ourselves with natural beauty. It may be an instinctive orientation that helps us do those things that serve our long-term interests. It is said that people will only protect what they love. Loving nature, biophilia, may be a crucial means of ensuring that we protect nature, and hence maintain the conditions for our own flourishing.

The future whispers while the present shouts. Desires demand immediate satisfaction, and amidst their constant shouting the voice of reason may be difficult to hear. Modern media and technology stimulate endless desire; they make it all too easy to make bad choices. As a passion, biophilia can compete with other passions and desires on their own turf. As such, it is a useful ally to reason, whose capacity to perceive long-term interest is notable but whose power to motivate action is negligible. Environmental activists consistently identify their love of nature as a crucial motivation for their public service. With this in mind, cultivating biophilia may be one of the best means of furthering sustainability.

A society that does not love and revere nature may not be able to discipline itself sufficiently to become sustainable. The temptation to overexploit resources for selfish, short-term gains may be too great. A race of Vulcans (the hyper-logical species featured on *Star Trek*) would undoubtedly live sustainably in order to serve their long-term interests. As a species motivated by emotions, however, humans need to make good use of biophilia to ensure that they do the right thing. Biophilic desire may be necessary today to counteract the short-term pleasures and comforts that undermine our long-term welfare.

This is not to say that love of nature is sufficient for sustainability. Biophilia, on its own, does not show us what actions are necessary to sustain a particular ecosystem, community, society, or global system. For that we require reason, science, critical and constructive dialogue, and a good deal of trial and error. In turn, biophilia is not the only passion at our disposal. The love of children is a powerful motivator for practitioners of sustainability. And our sense of duty to future generations and to people distant in social or geographic space, duties grounded in a love of humankind, is also crucial. Fortunately, biophilia does not and need not work alone. Our love of nature can and should complement parental love, humanitarian love, our sense of moral duty, and rational self-interest to motivate the practice of sustainability.

Of course, an abiding love of nature does not guarantee concern for human beings distant in time and space. It does not even necessitate concern for family and friends. A misanthrope is someone who dislikes, has contempt for, or even hates people. Misanthropes might well love and revere (non-human) nature. Indeed, some "nature lovers" have developed misanthropic tendencies precisely because that which they most love – the pristine, natural world – has been despoiled and destroyed by human hands. They painfully experience the loss of species and the degradation of nature, and cannot abide the greed, ignorance, and shortsightedness of fellow humans who commit these wrongs. Nature-loving misanthropes come to hate, or at least deeply mistrust and resent, those who hurt the object of their love. And they tend to lump all human beings into a single category of wrongdoers, a "cancer" afflicting the natural world. To be sure, the rapid growth of human population across the globe may bear some resemblance to the growth of cancer cells in an organism.[28] But it is difficult to cooperate with others in the service of protecting nature when you are filled with fear, hate, or disdain for them. Misanthropic biophilia is not a recipe for building sustainable societies.

Sustainability requires the wedding of ecological caretaking with a temporally and geographically expanded concern for fellow humans. These should be mutually supportive endeavors. After all, human beings are a part of nature. Biophilic and humanistic tendencies can form a powerful alliance. Together, they constitute a strong force for the fostering of sustainable societies.

Environmental health

The promotion of environmental health links the caretaking of nature with concern for human welfare. Environmental factors – that is to say, pollution, lack of sanitation, and other forms of degradation of the natural world – account directly for at least a quarter and perhaps more than a third of human illnesses and diseases.[29] A half-century ago, Rachel Carson observed that, for the first time in history, every person on the planet had chemical toxins in their bodies. Today, our bodies are contaminated with scores of synthetic chemicals and metals. These include toxic ingredients from rocket fuel and flame retardants, industrial chemicals such as polychlorinated biphenyls (PCBs), dioxins, and other organochlorines, phthalates and bisphenol A (BPA, used in plastics), lead, mercury, cadmium and other heavy metals, and pesticides including residues of DDT (even though this pesticide has been banned for decades in many countries). These contaminants, which typically accumulate in the fatty tissues of our bodies, can have severe health implications, especially for young children and the fetuses of pregnant women.[30] They do their damage as neurotoxins, destroying or damaging brain cells, endocrine or hormone disruptors, immune system suppressors, and carcinogens or cancer-causing agents.

Living sustainably requires the conservation of natural resources. Chief among these resources is the environment's capacity to absorb the wastes produced by human societies without becoming degraded or toxic. It is often said that the "solution to pollution is dilution." To some extent this is true, at least with regards to most forms of environmental pollution. What typically transforms a substance into a toxin is its level of concentration. That is to say, in a sufficiently diluted form, most chemicals are not harmful to life. When concentrated (and combined), however, they may become deadly toxins. And many of the tens of thousands of synthetic chemicals that are produced and released into the environment today prove toxic in very small amounts.

Insufficiently diluted, even natural waste products become environmental hazards. For example, when animals range over large areas their excrement becomes widely distributed. As such, it serves a very useful purpose: it reintroduces nitrogen and phosphorous to the soil and enhances the land's capacity to nourish plant life. Waste from fauna effectively becomes food for flora. The plants that benefit from enriched soils can then be eaten by herbivorous animals, which can subsequently be consumed by carnivorous animals. This is how self-sustaining, closed-loop systems are maintained in nature. It is the circle of life – recycling at its best. Turning waste into food is what nature is all about.

Turning waste materials (from crops and livestock) into food is a key feature of permaculture. The word "permaculture" was coined to designate a form of (permanent) agriculture based on the careful observation of and participation in natural systems. The original aim was to create self-sustaining means of food production. But permaculture has come to include the ecological design of most if not all components of human communities, including home building and landscape restoration.

Unfortunately, contemporary agribusiness carried out by large corporations does not practice permaculture. When factory farms place large numbers of livestock or poultry in small spaces, their excrement ceases to become food for plant life (that in turn is eaten by herbivores). Instead, it becomes a potent toxin. Concentrated (or Confined) Animal Feeding Operations (CAFOs) may contain tens of thousands of animals, typically meat and dairy cows, hogs, turkeys, or chickens. The vast amounts of slurry (feces mixed with urine) produced in these mega-operations can scorch the earth, sully nearby waters, and fill the air with contaminants and stench.

Plants do not eat slurry. When otherwise beneficial elements and compounds are found in too high a concentration, they stop serving as nutrients and start acting as poisons. And when animal waste is not absorbed in nutrient form by plants, it leaches into underground or surface water or is absorbed into the air. In both cases, it becomes a source of pollution. Owing primarily to its high nitrogen and phosphorous content, and its pathogens such as *E. coli* bacteria, concentrated animal excrement can be more poisonous than industrial chemicals.

A full-grown hog excretes more than 15 pounds of waste per day. A cow excretes 25 pounds. A single factory farm holding thousands of animals may produce hundreds of thousands of pounds of slurry each day. Rather than serving as nutritious food within the web of life, this concentrated waste becomes a biocide. Factory farm slurry pools also emit large amounts of methane. And when slurry is spread on fields, nitrous oxide is emitted. Methane and nitrous oxide are very potent greenhouse gases, and nitrous oxide also degrades the stratospheric ozone layer that shields the earth from ultraviolet radiation. CAFOs threaten environmental health by concentrating naturally occurring substances to the point of toxicity.

Like methane and nitrous oxide, carbon dioxide is a naturally occurring substance. Without it, life as we know it would not be possible on the planet. Carbon dioxide is a waste product of the metabolism of many forms of life, as well as the burning of organic matter and fossil fuels. Unfortunately, its concentration in the atmosphere and in the seas has risen to dangerous proportions, producing climate change and ocean acidification. Today, owing to its levels of concentration, carbon dioxide is one of the most dangerous pollutants on the planet.

Excess carbon dioxide is toxic in two different ways. First, it causes heat pollution. It increases the temperature of the planet, making certain environments too warm to be hospitable to their native species. Plants and animals that cannot quickly migrate to colder lands or waters when their own habitats warm may perish. Many corals, for example, are bleaching and dying owing to warming waters. Second, climate change chemically pollutes the seas. Much of the CO_2 that we put into the atmosphere does not stay there. Over 20 percent of CO_2 emissions is absorbed by the oceans. This changes the water chemistry, making it more acidic. The earth's oceans are now becoming acidic to the point of toxicity, as the calcium carbonate that forms the building material for the skeletons and shells of many marine organisms is depleted. Everything from microscopic plankton and krill to corals, sea urchins, clams, mussels, oysters, and shrimp – which collectively form the base of many oceanic food chains – struggle to develop and maintain their skeletons and protective coverings in increasingly acidic waters.

For the first 10,000 years of civilization, the solution to CO_2 pollution was dilution. The amount of carbon dioxide produced by human beings was small enough, and the atmosphere and oceans were large enough, that dilution rendered this crucial compound of life wholly unproblematic. Indeed, it would be wrong to label carbon dioxide a pollutant prior to the industrial age. Its concentration was simply too low to merit this designation. In the last 300 years, however, CO_2 levels in the atmosphere have risen more than 35 percent. Equally alarming increases are occurring in the oceans. Consequently, carbon dioxide has been formally recognized as a pollutant by many government agencies, including the US Environmental Protection Agency.

Resilient ecosystems depend upon healthy, unpolluted environments. While a healthy environment is the foundation for a resilient web of life, it would be misleading to imagine the planet's land, water, and air as wholly distinct or separate from the various species that exist within or upon these solid, liquid, and gaseous habitats. Soil, for instance, is produced by various geologic and climatic processes, such as the weathering and erosion of rock. But healthy soil is also produced and maintained through the interactions of scores of plants and animals. Take away these contributing species, and you undermine the health of soil. Forests, shrubs, grasses, and other plant life can only thrive if the soil in which they root is very much alive, benefiting from the activities of the bacteria, fungi, protozoa, nematodes, earthworms, mites, and other insects that help recycle organic matter and transform minerals into the vitamins, hormones, and nutrients essential to plant growth. At the end of the day, there is no way to separate a healthy environment from the welfare of its occupant organisms.

A healthy environment is not simply unpolluted space. It is a biologically enriched home. Healthy skies, lands, and waters are permeated with life. One might say that a healthy environment is both unpolluted and undepleted. It is characterized by the absence of toxins and the presence of biodiversity. The health of the environment is the collective health of the myriad plants and animals that inhabit it. Environmental health and ecological resilience are two sides of the same coin. For this reason, environmental caretaking entails more than pollution prevention. It is the concerted effort to maintain skies, lands, and waters both free of pollution and enriched with biodiversity. These two goods are mutually reinforcing. Pollution is a grave threat to the health of species and the resilience of ecosystems. And a biologically depleted planet would be a much more polluted one. The flora and fauna making up a wetland, for instance, can filter and absorb human-produced toxins before they enter the aquifer. Indeed, the planet as a whole depends on the diversity of flora and fauna to keep it from becoming toxic. The plants and animals that populate the earth maintain its chemical balance.

In the 1960s, a scientist named James Lovelock was working on a NASA effort to detect life on Mars. He developed a theory that the living and non-living parts of earth formed a complex, interactive system that operated like a single organism. He called his theory the "Gaia hypothesis," naming it after the Greek goddess of the earth, the mother of all life. Lovelock's idea was that the biosphere, atmosphere, hydrosphere (water), and pedosphere (the earth's crust) interact in a self-regulating fashion to maintain a planetary environment that is conducive to life. Through complex feedback loops, and without any conscious, overarching purpose, the living and non-living parts of the planet produce the relatively stable conditions – temperature, atmospheric composition, ocean salinity, and soil fertility – required for biological processes.

To illustrate the point, Lovelock (along with Andrew Watson) proposed a hypothetical model of an earth-like planet orbiting a sun-like star. The star's radiation is growing in intensity, potentially threatening life by overheating the planet. But this planet has the good fortune to be covered with two types of daisies: some with black petals and others with white petals. As the solar radiation increases, the daisies with black petals, absorbing much more of the sunlight than the white variety, overheat and wilt. White daisies do comparatively better, and soon much of the planet is blanketed with them. But now the whitened surface of Daisyworld reflects most of the solar radiation back into space, effectively cooling the planet. Soon it gets too cold for the white daisies to flourish, and the black-petalled variety, which absorb more heat, slowly regains its lost territory. Now the black-and-white dappled planet starts to warm, and the process repeats itself. Lovelock's Daisyworld illustrates how living things (flowers) can interact with inorganic processes (solar radiation) to maintain a planetary habitat in a state of equilibrium that is conducive to life.

The Gaia hypothesis, like any scientific claim, has been subject to much debate. Lovelock joined Lynn Margulis, a microbiologist, to defend and further substantiate his theory. Margulis insisted that the planet should not be understood as a living organism and Gaia should not be personified as a conscious being. Rather, Gaia is an emergent property arising from the interactions of organisms and their environments. Still, from a certain perspective, that is, from outer space, the earth appears as if it were alive. Like any other living organism, it maintains itself in relative equilibrium by way of the interaction of its component parts. It is a self-regulating system, one that adapts in order to sustain itself.

Gaia is resilient. Over the eons of evolution, the web of life has suffered many shocks, and has always bounced back. The stark reality is that the chief threat to Gaia today comes from its own children: the human species. To acknowledge this is not to condone misanthropy but to underline the urgency of building sustainable societies: interdependent communities that are economically prosperous, socially empowered, and culturally creative. Only in this way can our species become responsible contributors to the dynamic, resilient ecologies that make the earth an enduring home.

Inquire and Explore

1. How much of the natural world is threatened today, and what can you do to conserve it?
2. How can you experience biophilia?
3. Should we be optimistic or pessimistic about preserving biodiversity and the web of life?

* * * * * *

- Learn about environmental pollution and how it can be prevented.
- Get involved in conserving nature and creating sustainable landscapes through citizen science projects.

 For web resources and more information on these topics, please visit conservationandcreativity.net

4

Technological Solutions and Their Problems

Built environments and production systems have made the earth into a human world. Built environments include all of the constructed dwellings, industrial and commercial buildings, walkways, bikeways, and roadways, energy delivery systems, sewer and waste disposal systems, and other forms of infrastructure that provide humans with places to live, work, recreate, and travel. About half of all materials extracted from the earth go into the construction and maintenance of buildings. Buildings also consume more than 40 percent of all energy generated and produce more than 40 percent of all carbon dioxide emissions.[1] Production systems are the technological means we employ to utilize natural resources and human services. They encompass the resource extraction, manufacturing, energy, transportation, and agricultural sectors of the economy. They also include the information and media, healthcare, education, retail sales, financial, and other service sectors. Production systems process virtually all the resources that we consume and they generate most of our waste.

Built environments and production systems create and comprise the material components of civilization. The way these material components are generated and maintained currently constitutes the most prominent threat to sustainability. Unless the technologies employed in our built environments and production systems become cleaner and greener, the goal of creating sustainable societies will elude us.

According to a report by the United Nations Department of Economic and Social Affairs, investment in green technology of approximately US$2 trillion a year for the next three to four decades is required to avoid "a major planetary catastrophe." Its lead author concludes that:

> Business as usual is not an option...Even if we stop the global engines of growth now, resource depletion and pollution of our natural environment would continue because of existing production methods and consumption habits. Without drastic improvements in and diffusion of green technologies, we will not reverse the ongoing ecological destruction and secure a decent livelihood for all of humankind, now and in the future.[2]

To achieve sustainability on a global scale will require technological innovation. Much of that innovation must occur in sustainable design, also known as environmental design or green design. Sustainable design organizes and crafts the built environment, production systems and technological devices so they operate in an environmentally benign fashion.

Sustainability design may produce wholly new, narrowly targeted, and sophisticated technologies that gain efficiency and reduce waste. LEDs (light-emitting diodes) are a case in point. Replacing incandescent bulbs with highly efficient LED lighting systems in the next two decades could save almost as much electricity as is currently being used for lighting in homes.[3]

Sustainability often requires us to find the right tool for the right job. At times, the "right" tool, machine, or technique may be simple, small, inexpensive, non-polluting, and versatile. Brooms and rakes, for example, may be better choices for home owners than noisy, gas-powered leaf blowers. At times, the adaptation of an age-old discovery might be best, such as traditional methods of crop rotation and permaculture that preserve soil fertility and prevent erosion. In turn, we must acknowledge that not every problem has a technological solution. And we must remember that every technological solution creates new sets of problems.

In this chapter, we investigate both the solutions that technology provides and the problems associated with these innovations. Every technological intervention produces unintended consequences. Today we face unintended consequences of such scope that civilization itself is threatened. But there is reason for cautious hope. Becoming aware of the ways in which our invented tools, machines, and processes can come back to haunt us allows us to design smarter, more appropriate forms of technology.

Our Midas touch

The relationship of technology to sustainability was first brought to the public's attention in 1971, with the publication of Barry Commoner's *The Closing Circle: Nature, Man and Technology*. Commoner argued that centralized and polluting forms of industrial technology were the least appreciated and potentially most important factors in explaining the environmental crisis. Taking issue with the then prominent focus on overpopulation, Commoner insisted that technology was the greatest source of pollution and that inequitable forms of distribution (which went hand in hand with centralized technologies) rather than outright shortages of food were responsible for most famines. Industrial technology, with its synthetic and often toxic products, by-products, and waste materials, as well as the accompanying maldistribution of benefits and risks, was the chief threat to environmental and social welfare.

Paul Ehrlich, an ecologist, along with John Holdren, a physicist, argued that Commoner had left key variables out of the equation. They accused him of promoting a "one-dimensional ecology" with a singular focus on "faulty technology." Ehrlich and Holdren insisted that the environmental impact of any particular society was a product not only of technology but also of human numbers and standards of living. The way that we produce things does matter. But if human numbers were very small, the environmental impact of their activities would be quite small too, regardless of the technological mode of production.

Likewise, if the standard of living was very low such that people consumed very few natural resources, their environmental impact would also be low. The now-famous equation that Ehrlich and Holdren developed to capture these relationships was $I = P \times A \times T$. "I" refers to environmental impact. "P" stands for population. "A" refers to affluence, understood as the level of per capita consumption, typically measured by GDP. And "T" stands for technology.[4] To assess the depletion and degradation of natural resources caused by any society (I), one multiplies the number of consumers (P) by their level of consumption (A) and by the technological means (T) they employ to produce and distribute their goods and construct and maintain their built environments.

The $I = PAT$ formula has been much debated over the last three decades. Many have suggested the need to weigh one or another variable more heavily. Some have argued that the "T" part of the equation should be a denominator rather than a multiple.[5] The sustainability of a society depends upon the size of its population relative to its resource base multiplied by its level of natural resource consumption. But if clean, green, and efficient technology decreases environmental impact, even beyond the low levels of impact that might be achieved without the use of technology, then the P and A of any given society must be divided by its T. So the formula would become $I = PA/T$. Given the right technology, large, affluent societies may still be sustainable.

Conceptually speaking, technology is easily distinguished from population and affluence. However, a society's level of technology affects its fertility rate, the longevity of its population, its level of consumption, and the sorts of things it builds and consumes. Technology can increase or decrease human numbers, stimulate overconsumption and waste, or foster efficiency and resource conservation. It can be a major cause of pollution, or provide a means for cleaner, less toxic production systems and built environments. Indeed, the most technologically advanced nations have managed to greatly diminish many forms of water, air, and land pollution – such as that caused by heavy metals, sulfur dioxide, sewage contamination, soot, and other particulate matter. At the same time, the most technologically advanced nations remain the most polluting in terms of carbon dioxide emissions and many forms of hazardous waste. One thing is clear: no account of sustainability can afford to ignore the impact of technology.

Human beings distinguish themselves from the rest of nature by their technology. Chimpanzees use long sticks to extract termites from their nests, and Egyptian vultures employ rocks to break open eggs. Indeed, many animals utilize rudimentary tools. But only humanity has truly reshaped its world – and itself as a species – by way of technology.

The forerunners to *Homo sapiens* were the early hominins. They employed technology more than 2 million years ago in the form of primitive hand axes. A hand axe was simply a piece of stone, often flint, which had been shaped by blows of a harder, less flaky stone to gain a sharp edge. This basic, handheld, multipurpose tool could be used for chopping up wood or food, breaking bones to retrieve nutritious marrow, scraping hides, cracking nuts, digging in the soil for roots, and making other primitive tools of wood or bone. For the most part, the hand axe did not change its form for 2 million years. Eventually, methods of polishing the handheld stones made them more efficient and durable. In turn, the stones were attached to wooden or bone handles, which made them useful for clearing land and digging furrows, contributing to the development of agriculture in the Neolithic period beginning about 10,000 BCE.

By this time, fire had already been in use for about a million years, mostly for cooking food. Around 8000 BCE, fire was harnessed for metallurgical purposes. Metals were smelted from ores, and primitive copper axes and other tools were crafted in forges. Tools proliferated in this Neolithic period, allowing for vast improvements in agriculture, housing, and craftsmanship. The trading of tools, and goods made from tools, also greatly increased. Economic exchange likely accelerated the development and spread of technology, a function it still serves today.

Not infrequently, technological transformations have led to the destruction of societies. In his best-selling book *Collapse: How Societies Choose to Fail or Succeed* (2005), Jared Diamond observes that ancient societies collapsed for any number of reasons, including military conquest, loss of trading partners, and environmental factors.[6] The chief environmental causes were deforestation and habitat destruction; soil erosion, fertility loss, and salinization;

overuse and overconsumption of natural resources such as water, game, and seafood; overpopulation; and the introduction of non-native species. The ancient societies of Easter Island, the Anasazi of southwestern North America, and the Maya of Central America all collapsed primarily owing to their degradation, overuse, or overconsumption of natural resources. In each case, the technological capacities of the society in question, coupled with population growth, likely produced the environmental destruction that led to collapse.

Diamond suggests that this pattern of unsustainable practices leading to societal collapse may replicate itself in contemporary times. But there will be new contributors to the crisis, including climate change, environmental pollution, energy shortages, and the excessive human appropriation of natural habitat. Notably, each of these contributors to collapse stems from our technological capacities.

In ancient Greek mythology, technology was seen as a gift from the gods. But it came with strings attached. The deity Prometheus had the honor of distributing to all the beasts of creation their unique capacities. He allowed his brother, Epimetheus, to carry out the task. Epimetheus duly gave the lion and tiger their powerful jaws and claws, the bears and elephants their brawn, the birds of the air their feathered wings, and the snake and the scorpion their venom. Epimetheus had fully depleted his gift bag at the end of a long day when he realized that one species, humankind, had been left empty-handed.

Learning of the oversight, Prometheus returned to Mount Olympus, the home of the Greek gods, and put an ember from the sacred fire in a fennel stalk. He then took the burning ember back down the mountain, and bestowed fire upon the human race.

When all-powerful Zeus, king of the Olympian gods, discovered the thievery, he chained Prometheus to a rock in the Caucasus Mountains, where a vulture ate out his liver each day. The recipients of the stolen fire also felt Zeus' wrath. He had a beautiful woman created to become the wife of Epimetheus, and presented the happy couple with a handsome box as a wedding gift. The beautiful woman was Pandora.

Unbeknownst to the newlyweds, Zeus had put every imaginable evil into the box – such as greed, hate, calamity, envy, adversity, the lust for power, hardship, and stupidity. When the newlyweds opened Zeus' enchanting gift, all the evils came streaming out, forever to infect the human race.

Armed with the Promethean gift of fire, humans set themselves apart from the rest of creation by way of their capacity to cook meals, light and warm their homes, harden pottery, forge metals into durable tools and weapons, and communicate over long distances with signals from mountaintops. In a word, human beings had acquired technology. Accordingly, they began to refashion their world according to their needs, desires, and dreams. But this power was accompanied by all manner of vice and misfortune. To employ technology is to open Pandora's box.

Human beings have become the dominant species on the planet, owing to their technological power. The myth of Prometheus illustrates that there is a price to be paid for such power. But there is no going back. What was loosed from Pandora's box cannot be retrieved. To forgo technology today would be to regress to a primitive way of life that few if any would be willing to tolerate and many would not survive. The average lifespan of our hominin forebears was less than half our own. We owe our long lives, and much of our health, nutrition, learning, and comfort to technological advances. Technology is an indispensable feature of contemporary human existence. For good and ill, it is here to stay.

The web of technology encircles the globe, and reaches into every aspect of our lives. To put this book into your hands, for example, required many sophisticated technologies, such as the internet, word-processing computers, printing presses, and distribution systems. And each of these technological innovations could only have been developed, and can only be maintained, owing to a vast array of other technologies that supply component parts and services. Even strong advocates of sustainability with great concern for technology's potential mischief daily rely upon its benefits.

The name "Prometheus" literally means forethought. To live up to our Promethean heritage, we would have to exploit natural resources and put our tools to use with foresight and efficiency. But as the ancient Greeks already knew, when human hands reshape the world through technological means, things do not always go according to plan. Indeed, the technological solutions ingeniously supplied by one generation typically saddle progeny with significant problems. Tools and machines designed for wholly benign purposes – such as automobiles, energy plants, factories, and agricultural implements – have had the unintended consequence of upsetting the balance of life on the planet by changing its climate, polluting its lands and waters, and decimating its biodiversity. Technology produces unintended consequences.[7]

The ancient Greeks had another myth that well illustrated the notion of unintended consequences. It was the tale of King Midas. Midas had been kind to a friend of the god Dionysus, and was rewarded for his charity by being granted one wish. As he was fond of his riches, and desirous for more, Midas wished that everything he touched be turned into gold. Dionysus tried to persuade the king to choose differently, but Midas, with visions of untold wealth, could not be dissuaded.

The wish was duly granted, and Midas walked about the land testing his magical powers. To his delight, everything he touched – stones, flowers, trees – turned into the purest gold. Before long, Midas had made himself the richest man alive. To celebrate, the king ordered his servants to set before him the grandest of feasts. He quickly sat down at the table and reached for some succulent grapes. But alas, they turned into rock-hard gold. He stabbed at roasted meat, but it too was transformed into metal. The finest wine brought to his lips became as indigestible as the cup that held it.

Though surrounded by untold riches, Midas realized he would soon starve to death. In a panic, he begged Dionysus to rescind the wish. The god took pity on the woeful king and deprived him of his golden touch. King Midas went on to live until a ripe, old age, relatively poor but also happier, and much more careful about what he wished for.

Side effects

Like the Promethean tale, the story of King Midas illustrates how our actions – and in particular those of a technical sort designed to give us mastery and control of the world – produce unintended consequences. And the magnitude of our technological power increases the gravity of these side effects. Consider two examples. Chemicals called chlorofluorocarbons (CFCs) were invented in 1928. Beginning in the 1950s, CFCs were brought into widespread use as coolants and propellants. They were used widely in refrigerators and aerosol cans, with CFC production doubling every seven years. More than a million tons of these chemicals were being produced annually by the mid-1980s. They were wonderful feats of technology, allowing us to refrigerate our food and medicines, preventing spoilage and disease. But CFCs had side effects.

When released from aerosol cans or leaked out of coolant systems, CFCs migrated to the stratosphere. Here, they destroyed ozone. The stratospheric ozone layer surrounding the earth protects organic life from harmful ultraviolet radiation (UV-B). Although scientific papers in the mid-1970s warned that the ozone in the upper atmosphere could be damaged by CFCs, only in 1987, with the Montreal Protocol, did the phasing out of the production of CFCs begin globally. Since then, the loss of ozone has been slowly decreasing, though its depletion still presents a severe threat to photosynthetic processes, aquatic plankton, and many of the earth's creatures, including human beings. We face higher rates of skin cancer, eye cataracts, and other ailments on account of the increased ultraviolet radiation reaching the surface of the planet. Our inheritance of stratospheric ozone came to us across the eons. It took the planet millions of years to produce it.[8] We tore apart this protective shield in 50 years, and planetary life will continue to suffer the negative impacts for the rest of this century, as CFCs already present in the atmosphere continue to destroy ozone.

Side effects also present themselves in hydraulic fracturing or "fracking." Energy corporations employ hydraulic fracturing to break apart rock formations up to 20,000 feet under the earth's surface, pumping water and chemicals deep into the earth to release oil or natural gas. The use of fracking to gain access to shale gas has been proposed as an antidote to diminishing oil supplies, and a means of avoiding the burning of more polluting fossil fuels such as coal. No doubt it serves these purposes. But there are unintended consequences, which may include the pollution of aquifers and groundwater, as well as the release of large amounts of methane from rock formations into the atmosphere, where it may more than offset any reduction in greenhouse gases that comes from shale gas replacing oil or coal as fuel. In turn, concerns have been raised that fracking may produce earthquakes and other seismic disruptions.

Perhaps most disturbing are the potential side effects of *geoengineering*: the technological alteration of the atmosphere, land, or oceans of the planet to decrease the negative effects of climate change. One of the most discussed geoengineering projects entails introducing large amounts of sulfur dioxide into the upper atmosphere to increase the amount of the sun's warming rays that is reflected back into space. Advocates believe that enhancing the earth's "stratospheric albedo" in this manner is necessary to slow global warming, given the inability of most nations to significantly reduce their carbon emissions. However, such geoengineering does nothing to actually reduce the level of greenhouse gases surrounding the planet and the accompanying acidification of its oceans. The damage to the oceans would continue, and as soon as the spraying of the sulfur aerosols into the atmosphere stopped, the warming of the planet would resume. Potential side effects include ozone depletion, increased acid rain, and the delay or diminishment of efforts to reduce greenhouse gas emissions.

Such geoengineering is at best a technological quick fix that temporarily abates global warming without addressing its underlying causes. At worst, its known side effects would cause significant harm, and that is to say nothing of unintended consequences that remain unknown at present.

When the Iroquois nations of America developed principles for decision making, they insisted that choices be made with the welfare of seven generations in mind. Given their technology, this was an appropriate level of foresight. Few if any of their choices could not be arrested or reversed within a few generations. If a course of action did not produce negative repercussions for seven generations, it was a pretty safe bet that unintended consequences would never occur.

We cannot make the same claim today. The power and scope of our technology has grown so dramatically that it will impact hundreds of human generations. We do not know the long-term effects of geoengineering. We do know that our vast production of greenhouse gases bequeaths a legacy that may extend thousands of years. We receive the benefits of fossil-fuel heated and cooled buildings, manufacturing, and transportation today. Our children, grandchildren, and great-grandchildren will suffer the repercussions of climate change.

The shadows cast by technology extend across space and time. As our know-how increases, the effects of our ingenuity – both intended and unintended – impact distant populations and progeny. To be sure, new products and processes will be invented in the future to address the unintended consequences of today's technology. Innovation always produces unintended consequences, and these side effects stimulate the development of technological solutions, which in turn produce new sets of problems to be solved. Is this a sustainable pattern of human progress, or a vicious cycle spiraling out of control?

Technological optimists acknowledge that technology will produce unintended consequences, but argue that solutions will be found in time. Historically speaking, they are on pretty firm ground. After all, necessity generally proves to be the mother of invention. Humankind has steadily invented new means to solve problems and exploit natural resources, making life more comfortable and efficient. Today, people living in the technologically most advanced societies spend only a small fraction of their working days securing the basic needs of life – food, water, clothing, shelter, and energy for light and warmth. The rest is given over to other pursuits and leisure. Given the ingenuity of our race and the economic and intellectual drivers that stimulate innovation, we can rest assured that future problems will generate a plethora of technological solutions. Whether these solutions will constitute cures worse than the disease is another question.

In any case, technological optimists argue, there is no viable alternative to further technological development. Given the benefits that modern life derives from technology – increased longevity, better health, greater economic well-being, rising levels of comfort, along with the accelerated development of science and culture – who would choose to go back to the brutal days of hacking out our lives with (or without) hand axes? To try to put a stop to innovation and the consequent development of human intelligence and welfare, or even to slow it down, technological optimists insist, would be nothing short of criminal.

Technological pessimists warn that our problems are growing markedly in size, scope, and complexity. With climate change as a prominent example, they insist that we no longer have the technological wherewithal to solve the problems that we are creating. Technological optimists, they suggest, resemble the character in a well-known joke who finds himself falling off the top of a skyscraper. Upon passing each floor on his accelerating descent to the hard pavement below he cheerily remarks: "So far, so good!" Historically speaking, technological optimists rightly insist that human beings have always invented their way out of problems. This is often called progress. But given the scale and scope of the problems we face today, there may be no second chances. When you are standing at the edge of a cliff, taking another step forward is not progress.

"In a time of great technological power," Wendell Berry observed, it would be imprudent to operate on the basis of the belief "that humans either know enough already, or can learn enough soon enough, to foresee and forestall any bad consequences of their use of that power."[9] Our species is smart enough to create a technologically sophisticated life. The question is, are we wise enough to sustain it?

The issue of technology's impact on population growth illustrates the complexity of the problems we face. For the first million years of human evolution, when technology played virtually no role, population was kept in balance with environmental resources. Our species was no different from any other in this regard. A relative growth in our numbers would be followed by a "die-back," as resources dwindled. Often disease would take its toll. The seesaw between population and resource supply oscillated throughout the long history of human development. Though the amplitude of the swings varied depending on local circumstances, on the whole human numbers were relatively stable. It is likely that the global human population never exceeded 15 million until the onset of the agricultural revolution.

Our species then saw its numbers grow markedly as technological innovation equipped us to cultivate crops, domesticate animals, fashion tools, improve medical treatments, construct housing, and make durable goods, such as cookware and pottery. Over the next eight millennia, the global population grew fifteen-fold. Still, 2,000 years ago there were no more than 200 million human beings living on the earth. Populations increased steadily over the next two millennia, even as war and disease periodically reduced numbers. Plagues killed hundreds of millions of people in Europe and China during the Dark Ages and again in the 1300s, sometimes diminishing local populations by half. Diseases carried by Europeans devastated the natives of the Americas, killing as many as nine out of ten indigenous people. Invasions and wars also decimated populations. Recovery from these sharp declines often took hundreds of years. By the 1600s, however, global population had reached a new height of 500 million. It would grow steadily in the centuries to follow.

Less than 200 years after the human population hit half a billion, it doubled. That landmark occurred at the turn of the nineteenth century. The 2 billion mark took only 123 more years to reach. The next doubling of the global population took less than 50 years. We hit the most recent milestone, moving from 6 billion to 7 billion people, in a dozen years.

Certain technologies have had the effect of decreasing human populations. Weapons have certainly taken their toll. And ever more crowded urban settings unleashed contagious diseases. But on the whole, human numbers have steadily increased owing to technology. The more than doubling of the world's population over the last half-century has not been the product of rising fertility rates, but declining death rates owing to technological innovations in health care, agriculture, and the built environment.

While much of the industrialized world is stable in population today, the developing world is still growing rapidly. Asia and Latin America, for instance, have more than doubled their populations in the last half-century. The continent of Africa has more than tripled its population in the same time period. Global populations are expected to reach over 10 billion by the end of this century, before leveling off and starting to decline. Currently, the world's population is growing by about 80 million people each year. Each and every day, food, clothing, housing, schooling, employment, and an abundance of other resources have to be found for an additional 225,000 people. How are all these people to be fed? Given the loss of existing arable land owing to desertification, and the decline in soil fertility owing to overuse and erosion, it appears unlikely that conventional agricultural methods will be able to meet the nutritional needs of a quickly growing human population.

Some argue that shifting to organic agriculture will save the day. Organic agriculture does not employ manufactured pesticides or fertilizers, plant growth regulators, antibiotics or growth hormones for livestock, or genetically modified organisms (GMOs). Organic agriculture has many benefits. It can yield better crop resistance to drought (owing to improved moisture retention in soils), better sequestration of carbon, lower out-of-pocket costs for farmers owing to the absence of pesticides and synthetic fertilizer purchases, lower energy consumption, improved soil fertility, reduced run-off, better erosion control, reduced nitrous oxide release from synthetic fertilizers, and potentially improved nutritional and health value of produce. Organic farming also consumes less fossil fuel than conventional agriculture. Indeed, every 2 acres of land switched from conventional to organic farming reduces fossil fuel use as much as taking a car off the road. There is no doubt that well-managed soils can efficiently convert carbon from a greenhouse gas into food and feed. Indeed, it is claimed that "regenerative" organic agriculture, if applied to all current croplands and pasture, would be capable of capturing or sequestering much if not all of the world's current carbon emissions.[10]

Under certain conditions, organic agriculture may also generate increased yields per acre.[11] But such increases in yield appear to be relatively modest, and may not be able to support predicted increases in human population. And many studies suggest that organic methods, owing primarily to the absence of synthetic nitrogen fertilizer, produce lower yields than conventional agriculture.[12]

In light of the need to produce more food, the American agronomist Norman Borloug (1914–2009) argued that agricultural technology could and should come to the rescue. He predicted that the world's food production might be doubled by way of the genetic engineering of plant species to make them more productive of nutrients and more resistant to disease, pests, and drought. To feed the earth's growing population "on a sustainable basis," Borloug insisted, required genetically engineered crops.[13]

Borloug has impressive credentials when it comes to agricultural productivity. His endorsement of biotechnology was made in a speech commemorating the thirtieth anniversary of his receipt of a Nobel Peace Prize for seminal contributions to the Green Revolution. The Green Revolution began in the 1940s and reached its height in the 1970s. Initiated by Borloug, it produced high-yielding, disease-resistant crops, such as wheat, by cross-breeding a variety of species. Its explicit goal was to feed a larger proportion of the quickly growing populations of the developing world.

The Green Revolution largely achieved this goal. An estimated billion people were spared from hunger or starvation by the increased productivity of agriculture. That is a tremendous feat. But there were unintended consequences, many of dire significance.

The high-yielding crops developed by Borloug and others required increased irrigation. Consequently, aquifers in areas transformed by the Green Revolution have been severely depleted, and in many areas can no longer sustain high-yield agriculture. The salinization of soils has worsened owing to heavy use of potassium fertilizer and increased irrigation, which allows salts in the subsoil to leach up into topsoil. The heightened use of (mostly fossil-fuel based) fertilizers and pesticides has been a financial hardship for subsistence farmers, while causing widespread contamination of lands and waters. In turn, intensive plowing and other agricultural practices have depleted topsoil. A third of all agricultural land, by some estimates, has been eroded in the last half-century to the point that it is no longer farmable.

The Green Revolution also stimulated the use of fossil-fueled machines and the growth of large-scale agribusiness, concentrating power and wealth in fewer hands while reducing the biotic diversity of agricultural crops and the number of small farms and farmers. Last, but not least, increased food production allowed for the growth of human populations in the developing world. While it enabled vast numbers of people to be fed that otherwise would have gone hungry, the Green Revolution also helped ensure that there would be vastly more people to feed.

Would genetically engineered crops do better than Green Revolution agriculture at avoiding harmful side effects? That is an open question. We do know that genetically engineered crops tend to increase the use of herbicides (which these crops are engineered to tolerate). And the increased use of herbicides has already stimulated the mutation of herbicide-resistant "superweeds."[14] In turn, the development and use of genetically engineered crops will do nothing to reverse, and will much more likely accelerate, monoculture cropping, further decreases in the diversity of agricultural plant species, the growth of large-scale agribusiness, and the decline of small farms and farmers.

Even if we acknowledge that technological developments have produced dangerous problems, and that engineered solutions will generate their own, possibly graver, problems, the wholesale rejection of technology is not a viable option. The stable or declining populations of the advanced industrial nations demonstrate how expanded education and economic opportunity – substantially based on technological development – is the best means of lowering fertility rates. While the technological development of built environments and production systems has given us human populations that arguably exceed the earth's carrying capacity, their eventual stabilization and decline will require the further spread of technological innovations, something that typically occurs in tandem with economic development.

Cautious hope

Even the most hopeful of technological optimists have not figured out how the increasing resource needs of 7–10 billion people can be sustainably satisfied on a planet suffering from ecological degradation, dwindling biodiversity, sizeable declines in key natural resources, and climate change. The difficulty is deepened when we recall that the wealth and welfare of the advanced industrial nations over the last two centuries – wealth and welfare that yielded stable populations – were gained through their exploitation of vast quantities of fossil fuels. Fossil fuels are a finite and diminishing resource. It is likely that we have already passed peak oil, the point where the supply of inexpensive oil on the planet can continue to meet ever-rising demand. And even if reserves of coal, oil, and natural gas prove sufficient to meet the needs of a growing and increasingly wealthy global population, the overwhelming scientific consensus is that exploiting these fossil fuels at current rates, let alone increasing rates, will accelerate climate change to a catastrophic degree.

To be overly optimistic in current circumstances is simply not to see the whole picture. At the same time, we would be severely hamstrung in our efforts to tackle such massive and pressing problems were we to spurn technological innovation. Climate change, to be sure, cannot be well addressed without significant technological development in the arena of renewable energy, such as wind, solar, and perhaps tidal and geothermal. But should we allow every technology that comes round the corner to be developed and employed? Are there some things we can do that we should not do? We can produce weapons of mass destruction – biological, chemical, and nuclear. But clearly we should not allow their unchecked development and proliferation. Who would want a next-door neighbor engineering biological weapons in his or her garage? Likewise, who would endorse the unrestricted experimentation with human genetic manipulation? Given the high costs and risks of modern technology – economic, ecological, and social – there is no reasonable argument for its unregulated development.

Likewise, to spurn technological innovation when it remains a crucial feature of any viable effort to enhance the sustainability of societies is to fail to meet the contemporary challenge. Clearly, we cannot forgo all technological development. To do so would require intolerable restrictions to our intellectual and economic freedoms. And it would strike from our hands the chance fully to participate in the development of sustainable societies.

In this light, the task at hand is carefully to examine each issue in its context, arriving at a prudential decision based on best-available evidence. We face the need – a desperate need – for sound judgment that allows us to develop and employ technology selectively. We do not have the option of wholly arresting the technological odyssey that has occupied our species for hundreds of thousands of years. But neither ought we to resign ourselves to a technological free-for-all.

"Our ability to change the face of the earth," Bill Mollison observes, "increases at a faster rate than our ability to foresee the consequence of change."[15] And even when we can foresee the effects of change, our ability successfully to grapple with its unintended consequences remains meager. A technological carte blanche, allowing everything that can be done to be done, is a recipe for disaster. We require intellectual and moral discipline, and functional political institutions, to avoid the fate of Midas.

Being technologically pessimistic or optimistic rather misses the point. It is not helpful to fear and spurn all technological change, for such change is inevitable and often beneficial. But it is dangerously naive to welcome all technology, as much of it may cause more harm than good. Perhaps the best option is to be cautiously hopeful. As opposed to optimism, which assumes that a better future is inevitable, hope is an attitude about the present. To be hopeful is to see the present as pregnant with possibilities. Choosing well among those possibilities, and working to ensure that the best choices come to be realized, is the work of sustainability.

The "precautionary principle" reflects this prudent engagement with possibility. One of the most frequently cited definitions of the precautionary principle, known as the Wingspread Statement, reads: "Where an activity raises threats of harm to the environment or human health, precautionary measures should be taken even if some cause and effect relationships are not fully established scientifically. In this context the proponent of an activity, rather than the public, bears the burden of proof."[16] In short, when the consequences of technology are uncertain and potentially harmful, one should err on the side of caution.

The cause-and-effect relationships addressed in the precautionary principle are the unintended consequences of technological endeavors. The full range and impact of these side effects, despite the most rigorous scientific oversight, cannot be determined with certainty. So, in the absence of certainty regarding potential harms, efforts to regulate technological development are in order. The second clause of the principle holds the developers of technology responsible for investigating and assessing the potential risks of their activities and safeguarding the environment and society from potential harm. The burden of proof lies with the developers of technology, not its potential victims.

To employ the precautionary principle is not to eliminate risk. That would be impossible, for to avoid any particular risk in life typically entails facing some other risk. To avoid the risk of ill health, for example, one must exercise regularly. But exercise presents its own risks of physical injury. The attempt to avoid all or even most risks guarantees only that one will not live well or fully. Risk and the good life are partners.

With this in mind, the precautionary principle entails the systematic and comparative assessment and evaluation of risk. And it proposes a means of accountability so that risks and benefits are equitably shared. Those who develop technology should not get most of the benefits when things go well while the public bears most of the harm when side effects come home to roost.

A large number of civic organizations, business corporations, national governments, and international agencies have formally adopted the precautionary principle. Arguably, its popularity largely reflects its vagueness. The principle can be invoked to reassure the public of the moral stature of an organization, business, or government agency without, at the same time, committing its adopter to any specific measures. To be sure, the precautionary principle does not spell out what level of risk is acceptable, or how risk should be mitigated. It is not an algorithm that can be mechanically applied, guaranteeing specific forms of restraint or even consistency among cases.[17] Like other moral principles, it provides a guide to action. It calls for an ancient virtue, practical wisdom, to be exercised, reflecting Cicero's conviction that precaution is better than cure. And the precautionary principle goes some distance to setting out how this conviction should impact our concrete practices.

E. F. Schumacher observed that "today man is far too clever to be able to survive without wisdom." How might practical wisdom guide our technological endeavors? Practical wisdom is a skillful way of grappling with an uncertain and changing world. This skillfulness gets developed through the hands-on education we call "experience." But practical wisdom is not in the business of preventing all mistakes from being made. Learning chiefly arises from error. The hands-on education that produces practical wisdom is typically gained in the "school of hard knocks." As the product of experience, practical wisdom develops when we make good use of our own mistakes, or even better, when we learn from the mistakes of others. That is why the practice of sustainability ought to be grounded in the study of history.

Practical wisdom is developed when we open ourselves to the learning that can occur in the wake of error-prone practice. Our knowledge, perspectives, and theoretical models never seamlessly map onto the world around us. So our actions will inevitably produce unintended consequences. Mistakes will be made. Hence the need for caution. Aware of the inherent limitations of all knowledge, perspectives, and models, the practically wise person expects the unexpected, and acts accordingly. To err is human. To err on the side of caution, and to learn from each error, is practical wisdom.

Adaptive management and appropriate technology

Whenever we develop and maintain built environments and production systems – industrial, commercial, residential, or agricultural – these enterprises impact the natural and social world. They may cause pollution, diminish or degrade natural resources, or reduce habitat for threatened species. They also produce social goods and ills, many of which will not be fairly distributed. Sustainability requires that built environments and production systems produce more goods than ills, that the social distribution of both benefits and risks is fair, and that the natural world's resilience is not diminished. One means of meeting this challenge, which cultivates practical wisdom in the service of sustainability, is adaptive management.

Adaptive management is a way of learning by doing. It treats our built environments and production systems as experiments. This is not to suggest that we ought to take a cavalier attitude toward the impacts that our technological endeavors have on the natural and social world. Indeed, it is precisely the potentially devastating consequences of our actions that necessitate an experimental approach. To manage our built environments and production systems adaptively means that we are ready and able to modify the way we build and maintain our places of work and residence and produce goods and services when problems arise.

Adaptive management was originally developed as a method of protecting ecological habitats and species through small-scale, cautious experimentation. Given the complexity of ecosystems, and the difficulty of predicting their responses to human intervention, it is very difficult if not impossible to design a conservation plan that can be implemented once and for all. A more humble, and more responsive approach is necessary. So adaptive management engages multiple stakeholders in an effort to balance social, economic, and environmental goods. It brings scientists and conservationists together with people who influence, and are influenced by, the health and resilience of local natural habitats, such as loggers, fishermen, real-estate developers, and residents. Any such effort will prove imperfect. But it is precisely its failures that provide the feedback needed to implement adaptive improvements.

Science advances when hypotheses are disconfirmed through experimentation. Likewise, adaptive management advances by way of mistakes made in small-scale efforts to protect local ecosystems while improving the lives and livelihoods of stakeholders. Planning and implementation are followed by feedback and evaluation, the revision of plans, and renewed efforts at implementation. A pragmatic, experimental approach produces data, and these data are collected and analyzed to allow for a better response to the unintended consequences of initial enterprises.[18]

Although adaptive management is designed for failure – that is, designed with the idea that small-scale failures are the chief mechanisms for learning and improvement – it remains inherently risk averse. It errs on the side of caution by limiting the scale of its experiments. It engages in experimentation that is "safe to fail."[19]

Adaptive management has known its share of frustrations and disappointments. Conflict among stakeholders is frequent, and consensus on values and processes remains elusive.[20] Owing to the costs of gathering and analyzing data and the inherent failures and uncertainties built in to the adaptive management process, conflict may be heightened. Some of this conflict will be beneficial, highlighting alternative perspectives and possibilities. But such conflict needs to be bounded, chiefly by political means, so as not to escalate and derail collaborative efforts. So politics, the art of compromise, is inherent to adaptive management.[21]

Of course, political compromises do not typically produce the optimal conditions for scientific experimentation. Often politics plays a larger role than it should in determining objectives, the scale and duration of experimental trials, and practical responses to feedback. So adaptive management always sits in an uneasy tension with political realities.

In most sustainability-related issues, trial-and-error learning is the best we can hope for. But a trial-and-error approach is often politically inexpedient. People crave certainty, and often demand the security of policies that are not subject to change. The problem with rigid policies is that they cannot adapt to unintended consequences. And unintended consequences are unavoidable because we live in an interdependent world. When you act within a web of relations, it is impossible to predict all the repercussions.

Technological solutions go wrong when they ignore the larger web of relations within which they operate. Wendell Berry writes that "A bad solution solves for a single purpose or goal, such as increased production. And it is typical of such solutions that they achieve stupendous increases in production at exorbitant biological and social costs."[22] As Berry was a farmer, we might revisit an example from agricultural production systems to illustrate his point.

In chapter 3, we examined the pollution produced by CAFOs. CAFOs are designed to maximize the production of meat, milk, or eggs at the least cost. For the most part, CAFO animals do not move about or otherwise expend energy. They eat the grain or forage provided in their enclosures in order to grow meat or produce milk or eggs as quickly as possible. Because animals are kept in such close quarters, infections and disease are endemic problems. Small doses of antibiotics are fed to CAFO animals regularly, mostly as a preventative measure and to stimulate growth.[23] A recent study by the US Food and Drug Administration documents that 70–80 percent of all antibiotics and related drugs sold in the United States – tens of billions of pounds annually – are administered to poultry and livestock. This widespread use of antibiotics has severe repercussions. One side effect is the rise in the virulence and numbers of antibiotic-resistant bacteria, so-called "super-bugs," which threaten human health and life. It is estimated that tens of thousands of Americans die each year from infections of drug-resistant strains of bacteria.[24]

Wendell Berry coined the phrase "solving for pattern" to identify how technologies can and should be developed with sustainability in mind. To solve for pattern is to address a particular need in a manner that sustains the network of processes and values within which it is embedded. Good technological solutions maintain the integrity of the web of relations that they impact. Since everything is connected in a web of relations, solving for pattern requires solving more than one problem at a time by paying attention to and reinforcing vital connections.

Raising livestock on pastureland rather than in CAFOs would significantly reduce the need for antibiotics. Pastureland does not require methane-generating slurries and sequesters more carbon dioxide than the fields that grow the grains used in CAFOs. The use of pastureland also decreases the use of fossil fuels (for fertilizer, pesticides, and farm machinery). And pastureland is less prone to erosion and produces far fewer pollutants than cropland, such as fertilizer and pesticide run-off into streams, lakes, and aquifers. Last but not least, growing crops to feed livestock in CAFOs reduces the amount of grain that is available to feed people. In some countries, livestock consumes five times as much grain as is consumed by human residents.[25] As meat consumption in many countries is too high for optimal health, diets lower in meat and higher in grain and vegetables would be beneficial. Such a change in diet would help ensure that forests and other lands rich in biodiversity are not converted to pastureland.

In sum, eating less meat, eggs, and dairy products and putting livestock from CAFOs onto pastureland would solve multiple problems within a connected human-ecological system by reducing soil erosion, fossil fuel use, the growth of antibiotic-resistant disease, and various forms of pollution, including greenhouse gas emissions, while improving the health of both humans and livestock.

Solving for pattern is not something that can be done abstractly, theoretically, or from afar. It must be done concretely by local stakeholders. "Problems must be solved in work and in place," Wendell Berry insists, "with particular knowledge, fidelity, and care, by people who will suffer the consequences of their mistakes. There is no theoretical or ideal practice."[26] In short, solving for pattern requires the use of practical wisdom to manage our problems adaptively, and to learn from our mistakes because we directly suffer from their consequences.

The technology developed from this adaptive process is often called "appropriate technology." Typically, appropriate technology is engineered to be energy efficient, limited in its resource use, non-toxic, relatively inexpensive, widely accessible, locally controlled, low impact, restricted in scope, and "safe to fail." If its side effects prove too dangerous – and no technological development is ever wholly free of side effects – then it can be terminated without further damage being done.

The development of appropriate technology is a facet of glocality, bringing concern for the global commons into relationship with connection, responsibility, and accountability to local communities and ecosystems. In his 1969 monograph, *Operating Manual for Spaceship Earth*, the visionary engineer and inventor Buckminster Fuller identified the planet as a "mechanical vehicle, just as is the automobile" and suggested that technological innovations would solve many if not most of our problems. In contrast, Wendell Berry argued that "The only true and effective 'operator's manual for spaceship Earth' is not a book that any human will ever write; it is hundreds of thousands of local cultures."[27] Appropriate technology is grounded in local cultures that manage their welfare adaptively with an eye to global challenges.

Technologies are things we invent or stumble upon that solve particular problems and address particular needs or wants. Our hominin forebears had a need to dig up, chop, break open, and cut things. The hand axe was a good technological solution. Today, our problems, needs, and wants are much more complex, and their technological solutions are much more powerful and dangerous. Our two-million-year experiment with technology has demonstrated its massive benefits. We have also experienced its frequent – and often catastrophic – side effects.

Notwithstanding the inevitability of unintended consequences, we cannot abandon the pursuit of technological progress. As Peter Allen observes: "The world will never stop changing, and therefore sustainability is really about the capacity to respond, to adapt and to retain our options. And the power to do this...lies in *creativity*. And in turn this is rooted in diversity, cultural richness, openness, and the will and ability to experiment and to take risks."[28] Sustainability celebrates appropriate technology, embraces innovation, and practices safe-to-fail experimentation. But it rejects the belief that tools, machines, and technical processes can solve all our problems.

At the heart of sustainability is the task of managing well the scale and speed of change. Technological innovation will play a crucial role in meeting this challenge. But technology is not a panacea. There are no silver bullets. And even well-aimed bullets ricochet, including those with silvery hues. At the end of the day, technology cannot identify our core values and relationships or relieve us of the obligation actively to conserve and enrich them.

Inquire and Explore

1. Is geoengineering a needed response to climate change, or is it a cure worse than the disease?
2. Will human ingenuity and technology deliver us a bright future, despite the earth's finite and diminishing resources?
3. Should you live according to the "precautionary principle" or the "pro-actionary principle"?

* * * * * *

- Investigate how sustainable design can create a more sustainable world.
- Learn about organic farming, regenerative agriculture, and restorative pastoralism.

 For web resources and more information on these topics, please visit conservationandcreativity.net

5
The Political and Legal Challenge

Sustainability is challenging not simply because the technical problems are significant and there is no shortage of uncertainties. It is challenging because there is no shortage of people who actively oppose the environmental, social, economic, and cultural changes that sustainability demands. Some disagree with the data, or the predictions, or the values upon which sustainability efforts are grounded. Others are simply comfortable with the status quo, and they desire a future where their considerable wants will continue to be met, even at the expense of other people's needs.

Winning over the hearts and minds of many of those who oppose sustainability may be achieved by good scholarship, patient argument, and convincing appeals to common values. Likely, many will drop their opposition, or become more active in their support, when the impacts of climate change, other environmental degradations, or social inequities become too devastating to ignore or deny. For the foreseeable future, however, hard-fought battles with opponents are inevitable. The battles to be fought cannot be won with arms and violence. Different means are required. To some extent, these battles can be won in the marketplace, as green products, services, and technologies outcompete their resource-depleting, waste-producing rivals. To some extent, they can be won in the realm of culture, as communities develop and share new values and practices. To a great extent, the battles have to be won on the campaign trail, in legislatures, and in the courts.

Sustainability requires that individuals do their share to protect and conserve the natural world and contribute to social welfare. But individual effort is not enough. Political effort and legal support are required. That is to say, we cannot effectively practice sustainability in the absence of collective action. And we cannot foster and maintain collective action in the absence of political and legal processes and institutions. Susan George writes that "We could enjoy a world that is clean, green and rich, providing a decent and dignified life for everyone on a healthy planet.... We have in our hands, right now, all the knowledge, tools and skills we need. The obstacles are not technical, practical or financial but political, intellectual and ideological."[1] Building sustainable societies requires changing how we think about and organize our collective lives.

Sustainability often seems like motherhood and apple pie, something universally appreciated and endorsed. But the political reality is that some see the practice of sustainability as eroding individual liberties and undermining the free market. Others deride sustainability as a convenient means for elites to pay lip service to environmental health and social empowerment while reinforcing the economic and political status quo. Undoubtedly, some advocates of sustainability and some of the things done under its name merit these accusations. Like every other ideal and practice, sustainability has no immunity from cooptation and corruption. Faced with this hard truth, the task at hand is to refine, enrich, and share our understanding of sustainability, while developing personal practices as well as political and legal mechanisms and institutions to further its pursuit.

We begin our assessment of these challenges with a discussion of political parties and policies, followed by an exploration of the relationship of sustainability politics to empowerment and transparency, the sharing of power and knowledge. A brief look at the impact of laws and the courts leads to an examination of the linkage between national security and sustainable societies. The chapter concludes with an examination of the non-governmental pursuit of sustainability, and the relationship of governance to self-reliant individuals and communities.

Political parties and policies

Politics is frequently defined as the art and science of government. Government concerns the processes and institutions that allow people to make binding collective decisions. So politics refers to the way people organize their collective lives. People organize themselves not simply to gain security and ensure survival, but to achieve other benefits. Politics helps us survive together; it also enables us to pursue the good life together.

In a world of scarce resources, where everyone cannot have everything he or she wants, conflicting interests will emerge. Politics helps us mitigate conflict in a world of scarce resources. It allows us to resolve our differences through persuasion and law rather than brute force and violence. In turn, politics helps foster cooperation.

In democratic nations, one of the primary mechanisms for citizens to participate in politics is by voting. Through the selection of political representatives in free and fair elections, citizens can influence policy, the crafting of law, and the control of governmental bodies. The vast majority of elected representatives today are members of political parties, organizations that promote a common set of interests or beliefs.

The first political parties that formed around sustainability-related concerns arose in the early 1970s in Australia, New Zealand, Belgium, Switzerland, and the United Kingdom. Later that decade, the German Green Party was organized. Known as "die Grünen," the German Green Party endorsed the non-violent pursuit of ecological values, social justice, and participatory democracy. The German Greens contested a national election in 1980, and won 27 seats in the national legislature or Bundestag in 1983.

Since then, green parties (under various names) have formed in many countries across the globe. The US Green Party formed in the mid-1980s, typically has well over a hundred representatives in elected office in any one year, mostly at local levels and in state legislatures, and claims more than three hundred thousand registered members. The Green League of Finland became the first party to win sufficient seats in a national government to participate in a ruling coalition in 1995, followed by the Italian (Federation of the) Greens (1996), the French Greens (1997), and the German Greens in 1998. In 2004, the leader of the Latvian Green Party became the country's prime minister (for a very brief period), the first Green Party politician to head a state.

Green candidates and parties, for the most part, have had limited success in gaining national political office, generally receiving less than 5 percent of the vote. At the same time, green parties now exist in most democratic countries, and they enjoy moderate success at municipal, county, provincial, state, and regional levels.

Politics is grounded in plurality, in the affairs of diverse groups of individuals with diverse interests. Securing public goods in the face of diverse ideologies and concerns entails compromise. Politics has rightfully been called the art of compromise. Green candidates and parties, no less than their mainstream rivals, have had to master this art. The German Greens eventually proved successful in changing national policy to forgo the development and use of nuclear energy, for example. But to achieve this goal, they found it necessary to support coalition governments and coalition policies that they otherwise would have preferred to reject. In the end, they agreed to support a large German military presence in Afghanistan after the 9/11 bombings. This political trade-off put the German Greens at odds with many domestic and international supporters who endorse non-violence generally or opposed this particular military campaign in particular.

Green party candidates often have their greatest impact not by becoming part of a governmental cabinet or a ruling coalition, or even by getting elected to office. Rather, they push mainstream candidates and parties to adopt more sustainability-oriented positions. To the extent that green candidates and parties strike a chord with the electorate, gain media attention, and participate in local, regional, and national debates, they can shift the platforms and policies of major, well-established parties and ruling governments. The impact of their efforts, in this regard, is rather difficult to measure. It is not well gauged by counting the number of representatives that green parties elect or the number of governments they form. Rather, their efforts find fruit in shaping public opinion, shifting the legislative agenda, and contributing to the policies and laws that enable and oblige citizens and governments to operate more sustainably.

Sustainability politics goes well beyond the efforts of candidates or parties that label themselves green. This is especially true in the United States, which traditionally operates as a two-party system. Here, green parties, like other third parties, remain peripheral. Yet sustainability finds its way into political platforms, policies, and legislation in the United States and other countries without strong green parties. Indeed, the vast majority of the environmental and sustainability policy and legislation adopted by national governments over the last half-century was the product of mainstream political parties of the center-left or center-right.

Sustainability-oriented policies and legislation may take the form of regulations that restrict the activities of citizens, businesses, and governmental bodies in order to protect natural resources, maintain health, and ensure social welfare. Examples of government agencies in the United States that issue regulations relating to environmental caretaking and conservation include the EPA, the Fish and Wildlife Service, the Bureau of Land Management, the Forest Service, and the National Park Service. These agencies determine the ways and means by which broad national laws are to be understood and enforced. Their regulations may stipulate, for instance, the amount and type of pollution that particular industries or machines (e.g., vehicles) are allowed to emit, the kind and amount of wildlife that can be hunted or harvested, the extent to which natural resources can be exploited, and the natural habitats one can and cannot disturb by various activities.

While regulations issued by governmental agencies may be restrictive to businesses and affect their bottom line, that does not mean regulations do not produce greater economic benefits, in addition to environmental and social welfare. A study by the US Office of Management and Budget (OMB) demonstrated that regulations issued by the Environmental Protection Agency over a ten-year period imposed costs to the US economy of more than US$30 billion. However, the economic benefits of these regulations over that same time period were estimated at 3–20 times as great.[2]

Sustainability-oriented policies and legislation also utilize financial incentives. Green technological innovation, such as the development of renewable forms of energy, may be stimulated through tax breaks, rebates, or other incentives to industries, entrepreneurs, or consumers. Green services and products, such as public transportation, recycling, family planning services, and the provision of other public goods may also be stimulated through subsidies or tax breaks to service agencies and businesses.

Some policies foster sustainability neither by restricting nor subsidizing activities. Rather, they engage in "choice architecture." In their book *Nudge: Improving Decisions about Health, Wealth, and Happiness,* Cass Sunstein and Richard Thaler observe that the way alternatives are presented to people can have a significant effect on what they choose.[3] For instance, schools can get students to eat better foods at their cafeterias by putting healthy selections at the front of the display, with the junk food placed in the back. Likewise, governments can make more organs available for lifesaving transplants if the consent of potential donors is presumed when drivers' licenses are applied for or renewed. In such cases, people are given explicit directions for how to opt out of the organ donor program. Just as relatively few people choose to opt in when given the choice, relatively few choose to opt out.

Choice architecture does not impinge on the liberty of individuals to make their own decisions. Rather, it structures the context in which choices are made, tilting people toward one alternative or the other. This tilting is grounded in human psychology, which is to say, the way people typically make decisions. In the context of uncertainty, for instance, people shy away from decision making altogether, and hence refrain from taking action that alters the status quo. That is why participation in organ donor programs is greatly increased when opting in constitutes the default position.

Through the architecture of choice, people can be nudged (rather than forced) to make healthier, more sustainable choices without diminishing their freedom. Industries and retailers spend enormous sums of money nudging consumers in the marketplace with advertising. The aim is to direct and increase consumption. The architecture of sustainable choices employs nudging by public agencies to produce public goods, such as health, social welfare, and environmental protection.

The architecture of choice has the benefit of being safe to fail. Nudges can be tried experimentally on a small scale, with successful efforts implemented more widely. While the architecture of choice is still a relatively minor player in the field of sustainability politics and law, it is gaining favor in an increasing number of countries.[4] These efforts demonstrate that the politics of sustainability is not simply a matter of voting, joining parties, issuing regulations, or promulgating laws that mandate or restrict certain practices. It also concerns the informing of choice, the cultivation of preferences, and the development of good habits. People can be encouraged, stimulated, and persuaded to act in ways that improve the quality and sustainability of their personal and collective lives.

Many people may not want to be nudged. But if the benefits of nudging are clear and the alternative consists of more restrictive regulations and laws, then the architecture of choice provides the greatest social and ecological good with the least infringement of individual liberty. Poorly done and misapplied, it may become a corrupt and manipulative form of paternalism. To avoid this charge, it must be combined with efforts to educate and inform those whose choices are being shaped. That is to say, the architecture of choice must educate and empower stakeholders by making transparent both its purpose and the means selected to achieve it.

Empowerment and transparency

One might define politics as the process by which individuals and groups accumulate, exercise, and distribute power. Politics addresses key questions related to power. What should power be exercised to achieve (what collective goods, such as order and justice, need to be secured)? Who should wield power and how should it be institutionally distributed (what are the branches of government)? And what are the appropriate limits to the exercise of power (how do we protect individual liberties and rights)?

Empowerment is the distribution and cultivation of power within an ever-wider population. To empower is to include individuals and groups that otherwise would be inactive, marginalized, or disenfranchised, such that they are better equipped to define and pursue both their individual interests and public goods. The most basic kind of democratic empowerment comes in the form of a universal franchise, the power to vote.

The politics of sustainability requires people – as citizens, producers, consumers, and residents – to take responsibility for their own lives and the welfare of their families, communities, natural environments, and nations. This personal and collective responsibility can only be fostered through empowerment.

In the early 1900s, Gifford Pinchot addressed the relationship between environmental protection and democratic empowerment. Pinchot's notion of conservation might be seen as a forerunner to sustainability. He wrote that "Conservation is the most democratic movement this country has known for a generation. It holds that people have not only the right, but the duty to control the use of the natural resources, which are the great sources of prosperity." Focused on the benefit of the many rather than the profit of a few, conservation had the goal of making the country "the best possible place to live in, both for us and for our descendants."[5]

Later efforts to define sustainability also underlined the crucial need for democratic empowerment. In 1987, the World Commission on Environment and Development (WCED) maintained that "the pursuit of sustainable development requires...a political system that secures effective citizen participation in decision making."[6] Of course, endorsements of empowerment do not in themselves foster democratic practices. Critics argued that the WCED report did not promote grassroots democracy and local initiatives. Rather, it exploited the environmental threat to highlight the need for more governmental action. The WCED report was titled *Our Common Future*. But whose future is really being secured, critics asked, if the power to address environmental problems remained centralized in the hands of governing officials?[7]

Democratic empowerment depends upon transparency. Transparency refers to the removal of obstacles to the free flow of information and knowledge. In his letter of January

6, 1816, to Charles Yancey, Thomas Jefferson wrote "If a nation expects to be ignorant and free, in a state of civilization, it expects what never was and never will be." Jefferson went on to say that no people can safely live with liberty and be in control of property "without information." He concluded that a free press and an educated public were crucial ingredients of liberty and democracy. Knowledge is power, and the right to know is a right to empowerment.

To make informed choices, citizens and consumers need access to relevant data. People cannot make informed choices about the candidates campaigning for political office, for example, if they do not have accurate information about their backgrounds, achievements, and platforms. Transparency in government also requires open parliamentary debate and legislative (roll call) votes, so that citizens (informed by various media) may learn how policy and law get made and how elected representatives participate in its making. In many countries, the principle of transparency in government also requires that legislators disclose information about campaign contributions and their interaction with lobbyists, as citizens have a right to know how legislators are influenced in making decisions. This information allows citizens to hold representatives accountable for their deeds.

Just prior to the collapse of the Wall Street stock market in 2008 that led to a global recession and a massive government bailout of banks and firms, the finance industry had over 2,900 lobbyists working for them in the halls of power. During the previous decade, it had paid out US$5 billion dollars in lobbying fees and campaign contributions, mostly to ensure lax regulation.[8] Increased transparency of such information would have allowed citizens to hold their government officials responsible for a sustainable economy.

Citizens also require open access to information about the security and health of their communities, workplaces, and environment. Rachel Carson's efforts in the 1960s made transparent the relationship between pesticide use and the decimation of birds and other species. Carson's investigations stimulated citizens to learn more about the chemicals that were quickly becoming part of their lives. But such learning proved difficult, if not impossible, in the absence of information made available by industries, businesses, and governmental agencies involved in the manufacture, use, and regulation of these chemicals.

Carson's efforts inspired citizens to pursue "right to know" legislation. This legislation mandated industries and agencies to provide and release data pertaining to the sorts and amounts of chemicals that were being manufactured and released into the environment. In turn, "right to know" legislation was expanded to include information regarding the safety of workplaces and the level and kind of chemical contamination of neighborhoods.

"Right to know" legislation is grounded upon the principle that those who bear the risks associated with degraded environments have a right to be fully informed about these risks. When this information is made available, citizens can ensure that the health of the environment and the risks they face in their communities and workplaces become issues of debate and action. Significant reductions in the emissions of toxins, up to 40 percent, followed the passage of "right to know" legislation in the United States. This occurred not because the law required reductions but simply because the public was now informed, and could hold chemical manufacturers and emitters responsible.[9]

Transparency requires making information widely available. Often, however, the data most needed have not yet been gathered or analyzed. Only that which can be measured, it is said, can be achieved. To practice sustainability, progress toward clear goals must be measurable. The gathering, analyzing, and sharing of data – known as sustainability indicators, sustainability metrics, or sustainability indices – ensure that clear goals can be articulated and progress toward them can be measured.

Sustainability indicators include efforts to assess benchmarks. A benchmark designates where a community, business, or nation currently stands in relation to specific goals. It is a point of reference known as a descriptive indicator. Once benchmarks are established, one may then measure progress. Measuring the success of efforts to move beyond a benchmark provides a performance or efficiency indicator.

For instance, a municipal government may be interested in increasing the amount of recycling. A study reveals that currently 20 percent of household refuse is recycled, while 80 percent goes to landfill. Having established a benchmark of 20 percent (and perhaps comparing this benchmark to what is achieved by other towns), the city council may now set a reasonable goal and a timetable for increasing its recycling rate. As progress toward this goal is monitored, performance indicators will help determine if the means chosen are working effectively.

Cities, states, and nations pursuing sustainability have developed many (descriptive and performance) indicators of environmental and social conditions and trends. These include per capita solid waste generated; per capita water and electricity consumed; recycling rates; percentage of days with acceptable air quality; percentage of population using public or alternative transportation; number of green public buildings; total acreage in parks and non-developed green space; percentage of families living below the poverty line; availability of affordable housing; percentage of unemployment; percentage of population registered to vote and participating in local, state, and federal elections; and percentage of locally owned and operated businesses.

Numerous governmental and non-governmental agencies and organizations, such as the International Sustainability Indicators Network, promote the development of sustainability indicators. Comparative data help those beginning a project to see what has been achieved by other businesses, cities, states, or regions. In turn, various sustainability benchmarking and performance indicators have been created to aggregate data at the national level.

The 2014 Environmental Performance Index (EPI), for instance, examines 20 different indicators within the nine broad policy areas of climate and energy, biodiversity and habitat, fisheries, forests, agriculture, water resources, water and sanitation, air quality, and health impacts. Surveying 178 countries, EPI measurements place Switzerland as the best performer, with Australia in second place; the United Kingdom is in 12th place, Canada is in 24th place, and the United States is in 33rd place.[10] Alternatively, the 2014 Sustainable Societies Index (SSI) ranks 151 countries employing 21 indicators, including gender equality, income distribution, education, employment, and public debt. Its indicators fall into the three categories of human well-being, environmental well-being, and economic well-being. In these three categories, which are not combined for an aggregate score, the respective rankings vary for Australia (44, 145, 7), Canada (23, 136, 65) the United Kingdom (38, 103, 54) and the United States (40, 139, and 96).[11] Obviously, different indicators yield different results. The Happy Planet Index, in turn, measures the well-being of people and their environmental impact by integrating indicators of life satisfaction, life expectancy, and the per capita environmental impact.[12] It has been proposed as an alternative to GDP as a measurement of national success.

An increasingly prominent approach to sustainability indicators is ecological footprint analysis (EFA). EFA aggregates a large amount of data concerning levels of resource consumption and the production of waste (including greenhouse gases and other forms of pollution) to yield a single quantitative measure, the ecological footprint, which is measured in land area (typically acres, hectares, or square miles).

EFA calculates the land area that is required to sustain a particular human population, given the kinds and amounts of things it consumes and discards. A decreasing ecological footprint means that a population's consumption and waste are becoming more sustainable. A large ecological footprint indicates that a population is engaged in "ecological overshoot" and is "running a negative ecological balance."

Studies reveal that ecological footprints associated with built space, cropland, and fisheries are not sustainable. That is to say, at current levels the construction of towns and cities, agricultural production, and the use of marine and inland fisheries are creating a global deficit. We are depleting habitats and species, fertile soil and arable land, and fish and seafood on the planet faster than they are regenerating. Three-quarters of the oceans' fisheries, for example, are either being exploited beyond capacity or are currently recovering from overfishing.

When these and other indicators are aggregated, our species appears to be exceeding by a half the ecological limits of the planet. Put differently, it takes the earth 18 months to regenerate and recuperate from one year of human use. Ecologically speaking, we are living on borrowed time. To sustain our current rates of consumption and waste, we would need to grow the planet by two more continents and as many oceans. By 2030, we would need a whole other planet. And if everyone on the earth were to consume and waste as much as the residents of the most developed nations, we would require the resources of four new planets. These are sobering facts. The good news is many nations' ecological footprints are declining, albeit slowly, as can be seen in the data collected and analyzed by the Global Footprint Network.[13]

The science of estimating ecological footprints and measuring and analyzing other sustainability indicators is still in its infancy. There are many disagreements about which data to collect, the accuracy of the data collected, and the appropriateness of basic assumptions, forms of analysis, comparisons, and rankings. While EFA efforts have produced varied results, almost all conclude that we have exceeded the carrying capacity of the planet by 30–50 percent.[14] Whatever the precise figure – and it will inevitably be a changing one – clearly our species is not living sustainably.

EFAs demonstrate that levels of consumption and waste will have to decrease significantly over the coming decades. Alternatively, we might continue on our current trajectory of resource consumption and waste production and instead decrease markedly the number of human beings on the planet. However, we would have to decrease human numbers by a third or more to achieve a sustainable planet. All indications suggest that the global population is on a path to increase over the next century – by three billion people. As the chances of adding new continents and oceans to the earth are nil and there are no other habitable planets in the vicinity, EFAs demonstrate that we are going to have to redesign, reduce, reuse, and recycle much more of what we take, make, and waste. The politics of sustainability is tied to our capacity to know where we are, where we want to go, and whether we are making any progress in that direction. None of these crucial questions can be answered without gathering and analyzing data about the environmental, social, and economic impact of our choices and activities. Only when this information is made transparent, that is to say, understandable and widely available, will citizens, policy makers, and legislators be truly empowered to act. What Jefferson said about freedom applies equally to sustainability: no society can be both ignorant and sustainable.

National and international law

One of the most effective mechanisms for pursuing sustainability at the national level is legislation. Most of the legislation that is widely regarded as pertaining to sustainability is known as environmental law. Environmental law at a national or domestic level is primarily focused on the control and remediation of the pollution of air, land, or water and the conservation and management of natural resources, including forests, fisheries, wildlife, parks, and other wild or natural spaces. In most countries, environmental laws exist at the national or federal level, as well as at the level of states or provinces and municipalities. At the national level, these laws often take the form of statutes or acts passed by a national congress or assembly of legislators.

In the United States, for example, there are more than thirty statutes that fall under the rubric of environmental law, including the Migratory Bird Treaty Act (1918), the Clean Air Act (1970), the Clean Water Act (1972), the Endangered Species Act (1973), the Emergency Planning and Community Right-to-Know Act (1986), and the National Environmental Policy Act, which in 1970 mandated that environmental concerns be taken into account whenever federal agencies make decisions or take action. The EPA was created later the same year to write and enforce regulations pertaining to federal environmental laws while conducting research, assessments, and education.

Beginning in the 1970s, the United States passed many of the most advanced environmental protection laws in existence. Over the ensuing decades, both industrialized and developing nations enacted a plethora of environmental laws. Some of these were more comprehensive and stringent than the early US legislation, and some grapple with issues that have still to be addressed in the United States. The 2008 Climate Change Act of the United Kingdom, for example, sets a goal for decreasing greenhouse gas emissions by 80 percent of its 1990 baseline by 2050. The Resource Management Act of New Zealand, in turn, is specifically written to "promote the sustainable management of natural and physical resources" so that communities can provide for their health, safety, and social, economic, and cultural well-being while safeguarding ecosystems and "the reasonably foreseeable needs of future generations."[15] Likewise, a number of countries have advanced environmental protection by amending their national constitutions. Here, conservation and sustainability are not only promoted through legislative acts in statutes but have been incorporated in the most fundamental laws of the land.

For example, the Constitution of Poland (Article 74) states that "Public authorities shall pursue policies ensuring the ecological security of current and future generations"; that "Everyone shall have the right to be informed of the quality of the environment and its protection"; and that "Public authorities shall support the activities of citizens to protect and improve the quality of the environment."[16] The Basic Law of the Federal Republic of Germany (Article 20a) reads: "Mindful also of its responsibility toward future generations, the state shall protect the natural bases of life by legislation and, in accordance with law and justice, by executive and judicial action, all within the framework of the constitutional order."[17] The Venezuelan Constitution (Article 127) stipulates that "It is the right and duty of each generation to protect and maintain the environment for its own benefit and that of the world of the future. Everyone has the right, individually and collectively, to enjoy a safe, healthful and ecologically balanced life and environment."[18] France's constitutional Charter for the Environment observes that "The safeguarding of the environment is a goal to be pursued in the same way as the other fundamental interests of the Nation." It explicitly endorses sustainable development as the mandate to "reconcile the protection and enhancement of the environment with economic development and social progress."[19]

The constitutional enshrinement of environmental protection and sustainability principles is widespread.[20] Relatively few constitutional principles, however, gain full administrative support. Governmental agencies specifically tasked with the enforcement of environmental laws may not be created. And those agencies that are created often remain understaffed, have very limited budgets, and face well-organized opposition to their efforts from various sectors of business and society. This opposition frequently finds its way into the court system, where a judiciary must interpret environmental laws and determine their constitutionality, applicability, and need for enforcement.

For a century and a half prior to the formation of the United Nations in 1945, international law relating to the environment was limited to a handful of bilateral treaties signed by nations to conserve whales, seals, birds, fish, and seafood stocks, with one agreement limiting transboundary air pollution (the 1941 Trail Smelter agreement between the United States and Canada). From 1945 until 1972, international environmental law came of age, as numerous multilateral agreements reflecting growing popular and governmental concerns were reached.

The Stockholm Conference (1972) was the first government-sanctioned international meeting devoted to the environment. It resulted in the non-binding principles of the Stockholm Declaration, which asserted the sovereignty of states over natural resources as well as their responsibility to ensure that the exploitation of these resources did not cause damage to the environments of other states. The Stockholm Conference stimulated the formation of the United Nations Environmental Program (UNEP). Shortly after UNEP was formed, in 1983, the General Assembly of the UN convened the World Commission on Environment and Development. The World Commission issued its (Brundtland) Report a few years later, where the principles of sustainable development were first articulated.

Beginning with the 1972 Stockholm Conference, and increasingly so after the 1992 Rio Conference, the number of international environmental studies and reports grew markedly, as did the number of international environmental agreements. Multilateral Environmental Agreements (MEAs) are a form of "soft" law because they cannot be strictly enforced. They are said to "lack teeth." Such was the case with the 1997 Kyoto Protocol on Climate Change, negotiated under the auspices of the United Nations. Nearly 200 nations signed and ratified the protocol, which became active in 2005. About a fifth of signatory nations from the developed world agreed to specific reductions of greenhouse gases (more than 5 percent below their 1990s emission levels). Lacking enforcement capabilities, however, the agreement set insufficient targets for developed nations and failed to set any targets for developing countries, a number of which, such as China, India, and Brazil, have some of the largest and fastest-growing climate impacts. In turn, the Kyoto Protocol was not able to secure the participation of a number of key nations, including the United States. These factors contributed to the failure of the Kyoto Protocol to meet expectations. Indeed, global emissions have risen steadily since the treaty came into effect, though it is possible emissions would have risen more quickly without it.

Notwithstanding a history of failures, MEAs often translate into customary law with potent effects, and they influence the writing of national statutes and constitutions as well as sub-national legal frameworks. The Stockholm Declaration prompted nations to recognize the rights of future generations, and the legal status of future citizens has now been incorporated into a number of national constitutions and domestic statutory law. Likewise, the Rio (Conference) Declaration of 1992 thrust the precautionary principle onto the world stage.[21] Following the Rio Declaration, the precautionary principle was incorporated into many national laws, and into the Constitution of the European Union with the adoption of the Lisbon Treaty in 2009.

Not all MEAs completely lack teeth. The Convention to Regulate International Trade in Endangered Species of Wild Fauna and Flora (CITES) subjects its 175 member states to restrictions on the international trade of selected species of plants and animals. Since its institution in 1973, CITES has effectively ensured that imports, exports, and re-exports of more than 30,000 species, which otherwise might be freely traded as live specimens, dried plants, or as animal parts such as elephant ivory and various kinds of fur, must be authorized through a licensing system.[22] CITES forbids its members from engaging in certain trade relations with non-complying parties and non-parties to the agreement.

Similarly, the Basel Convention on the Control of Transboundary Movements of Hazardous Wastes and their Disposal, which came into force in 1992, was designed to prevent developed states from dumping waste in developing countries for a fee. The European Union and 182 nations which are parties to the Convention agree to abide by strict conditions on the import and export of designated wastes. They also adhere to requirements for giving notice, gaining consent, and tracking the movement of wastes across national boundaries. As the sole legal instrument addressing the global movement and safe management of hazardous substances and other wastes, the Basel Convention has been a powerful mechanism for environmental protection.[23]

One of the most successful MEAs is the Montreal Protocol, which was designed to monitor and phase out the production and use of substances that contribute to the destruction of stratospheric ozone. Adopted by the European Union and 196 states, the Montreal Protocol is the first universally ratified treaty in the history of the United Nations. It has been widely hailed as the world's most successful environmental treaty, and perhaps the most successful international agreement of any sort. Since its coming into effect in 1989, the Montreal Protocol and its subsequent revisions have largely ended the production and use of chlorofluorocarbons (CFCs) and other chemicals that destroy the protective layer of ozone in the atmosphere which shields organic life on the planet from ultraviolet solar radiation.

The enforcement and success of MEAs depend upon the willingness of member states to abide by the decisions and protocols that they have voluntarily adopted. Though sanctions may be levied against a party to an agreement that does not follow its rules, such measures are rare. Occasionally, the International Court of Justice (ICJ), created by the United Nations in 1945 and located in The Hague, Netherlands, adjudicates environmental disputes.[24] The ICJ, also known as the World Court, renders advisory opinions of legal import at the request of agencies of the United Nations or other parties, and renders binding verdicts on contentious disputes voluntarily submitted to it by nation-states. For instance, it adjudicated between Argentina and Uruguay regarding a large pulp mill that Uruguay built on the banks of the Uruguay River, which forms the international boundary between the two states. In its decision, the ICJ recognized environmental impact assessments as an obligation under international law. Here, transparency and sustainability metrics have gained a global legal foundation.

Unlike judicial decisions within nation-states, which are generally enforced by domestic police, the World Court does not have an enforcement mechanism at its disposal. In theory, the Security Council of the United Nations could enforce World Court decisions. In fact, this has never occurred. So a national government that has voluntarily become party to a case before the ICJ might refuse to comply with a ruling. But such non-compliance by a voluntary party would be publicly embarrassing and politically costly. More likely, governments unwilling to comply with a potentially unfavorable decision simply would not volunteer to submit a dispute.

The international politics of climate change took a new turn with the "Paris Agreement" of 2015. This MEA, known as COP21, is the most recent accord reached under the auspices of the ongoing United Nations Framework Convention on Climate Change. Ban Ki-moon, the secretary general of the United Nations, called the agreement a "monumental success for the planet and its people." Hans Joachim Schellnhuber, an environmental scientist and chairman of the German government's advisory committee on climate change, said the agreement marked "a turning point in the human enterprise, where the great transformation towards sustainability begins."[25]

Such enthusiasm stands in marked contrast to criticisms of the limitations or outright failure of previous UN-sponsored climate conventions, protocols, and conferences, such as Rio (1992), Kyoto (1997) and Copenhagen (2009). Unlike these earlier efforts, the Paris Agreement set transparent, mandated goals and gained the support of virtually all the world's nations, including the two biggest greenhouse gas emitters, China and the United States. The 195 signatory nations are required to monitor and provide public reports on their emissions levels and reductions, employing a universal accounting system. They are also required to participate in global conventions every five years, supplying updated plans to improve their efforts. While actual emission reductions remain voluntary, the monitoring and reporting structure is legally mandated. The expectation is that this transparency will produce a "name and shame" system, where domestic public expectations and global peer pressure will push nations down a low-carbon path.

The explicit goal of the agreement is to lower greenhouse gas emissions to a level that would not cause global temperatures to rise more than 2 °C (3.6 °F) above pre-industrial levels. This is the level of global warming that scientists believe might be tolerable; beyond that the most devastating effects of climate change would manifest themselves and prove irreversible. Importantly, we are already half way to that 2° limit. And some studies indicate that the carbon dioxide already present in the atmosphere will push us up to if not beyond this danger zone, as the heat-trapping effects of carbon dioxide molecules can persist for centuries. In any case, the explicit provisions of the Paris Agreement do not ensure that the 2° target will be met. At most, the provisions get us part way there. If nations make good on their modest voluntary pledges and go no further, the planet would likely warm by at least 3 °C.

Whether the Paris Agreement provides a historical turning point in the global effort to stabilize the planet's climate largely depends on whether the signatory nations meet and exceed their pledged reductions, the Intended National Determined Contributions (INDCs). The United States, for example, has committed to reduce emissions from 2005 levels by at least 26 percent by the year 2025. To meet this target, new legislation will likely be required to complement that already in existence, such as the Clean Power Plan put in place by President Obama in 2015. Under this EPA plan, carbon pollution coming from power plants would be reduced by 32 percent by 2030. However, the Clean Power Plan has been challenged in the courts, primarily by the fossil fuel industries, and in Congress, primarily by Republican legislators who deny the dangers of climate change. New legislation would likely face similar opposition.

One of the most important achievements of the Paris Agreement is the signal it sent to the business world. To meet and exceed their pledged reductions, nations will likely have to reduce or terminate fossil fuel subsidies, aggressively promote renewable energy, and preserve or increase forests and other carbon sinks. Quite likely, the adoption of a carbon tax will also be required to keep emissions within tolerable limits. Businesses, including the global financial and energy markets, will respond to these policy changes by investing in low-carbon technology, products, and services.

None of these political and legal achievements will be easy. There are many hurdles and uncertainties, and no guarantees of success. The fact that 2015 was the hottest year on record – surpassing the previous record of 2014 in terms of globally averaged temperatures over land and ocean surfaces and constituting the largest margin by which the annual global temperature record has been broken – should underline the gravity of the situation and the need for immediate, steady, and concerted action.[26] The Paris Agreement is a hopeful start – but by no means a victorious conclusion – to what will have to be a protracted global battle against devastating climate change.

Security, governance, and self-reliance

Historically, security has been the primary duty of national governments. They are mandated with protecting their citizens from domestic crime, acts of terrorism, and foreign invasion. Crime, terrorism, and war remain urgent concerns. Today, however, the security of citizens is increasingly put at risk by environmental degradation and resource depletion. National responses to these threats are often insufficient, as their resolution depends upon international cooperation.

Since the 1970s, scholars and analysts have been increasingly concerned with the security implications of environmental problems.[27] The vast majority of the world's large river systems, for instance, are shared by two or more nations. These shared rivers provide 40 percent of the freshwater resources to the world's population. As overuse and climate change causes these waterways to shrink or dry out, millions of people will likely face devastating shortages of fresh water. Social instability, civil strife and potentially violent conflict, so-called "resource wars," within and between bordering states may erupt. Initial flashpoints of conflict may include the Indus, Jordan, Nile, and Tigris-Euphrates rivers, all shared by multiple states.

The same threats surround the decline of other natural resources, such as oil, fisheries, forests, and agricultural land. In turn, an increasing number of states fail to provide for the most basic needs of their growing populations in the wake of environmental degradation and diminishing resources. Such "failed states" are prone to civil strife and insurgencies, and face large migrations as beleaguered populations flee across borders in search of food, shelter, and livelihoods. Neighboring states can then be drawn into the turmoil. These are all security issues.

The safety and welfare of citizens depend upon the safeguarding of environmental resources on both a national and global scale. Even the most militarily, technologically, and economically developed nations cannot insulate themselves from climate change and many other forms of environmental degradation. In a world characterized by ever broader and deeper connections, "ultimate security" requires global cooperation.[28]

As resource shortages combine with climate change to disrupt agriculture and raise demand beyond supply, food scarcity increasingly threatens national security. While grain harvests have tripled in the last half-century, the increase in production is largely a product of the extensive use of fossil fuels, primarily oil and natural gas, to run farm machinery and make fertilizers and pesticides. In turn, increasing amounts of this grain have been used to feed cattle and other livestock. So despite the vast increase in agricultural production, growing populations and growing levels (and types) of consumption have increased food scarcity. Future declines in oil production, which will lead to decreased supplies and higher oil prices, will worsen agricultural prospects. Indeed, agricultural productivity (the amount of grain produced per unit of land) is already flat or declining in many countries. The genetic engineering of crops appears unlikely to shift this pattern.

Reviewing the data on climate change, Lester Brown observes that wheat, rice, and corn yields could decline by as much as 10 percent for each degree (Celsius) increase in temperature. With global temperatures rising by as much as 6 °C within the century, the world may face massive agricultural shortages owing to drought and other climate disruptions. In turn, higher temperatures threaten glaciers. Melting snow and ice from Himalayan glaciers feed many of the world's major rivers, including the Indus, Ganges, Yellow, and Yangtze rivers. These rivers supply irrigation water to India and China, the world's top wheat and rice producers. A shrinking water supply will severely impact farming. The melting of glaciers, Brown concludes, presents "the most massive threat to food security humanity has ever faced."[29]

Global stocks of grains are at historic lows. As food becomes scarcer, its cost rises. In the last decade, food prices have risen steadily and hit two historic highs.[30] The recent tripling of corn, wheat, and soybean prices are cases in point. With billions across the globe already living well below the poverty line, higher food costs may prove disastrous. Food riots sparked by panicking people without the means to feed themselves and their families have already occurred in numerous countries.[31] Prospects for political unrest in food-scarce countries are worsened by the increasing foreign acquisition of agricultural land.

As demand in richer nations outruns domestic agricultural supply, governments and business corporations have begun buying up farmland in developing countries and exporting grains and other foods back to their home states. China, Saudi Arabia, and South Korea are at the forefront of this trend. They are purchasing or leasing agricultural land in places like Ethiopia, the Congo, the Sudan, Indonesia, and Vietnam. A number of the developing countries now selling land have historically faced famine, malnutrition, and chronic food shortages and have been dependent on regular food aid. Typically, the sale and lease of land to foreign investors is done without public input, without the involvement of local farming communities, and with little if any transparency. Foreign land acquisitions are likely to heighten food scarcity in developing nations, undermine democratic processes, and increase social and political instability.

Traditional threats to national security, such as the military power of large rival states, have arguably been superseded by environmental threats. In an age of climate change and rising oceans, for example, coastal defense cannot be secured by armed forces. Dikes, seawalls, and pumping stations will be required, and even more so international cooperation to sharply lower greenhouse gas emissions. With this in mind, Lester Brown argues that a Cabinet-level agency should be created in the United States, the Department of Global Security, to grapple with the primary dangers of the twenty-first century: failing states; climate change; food, water, and energy scarcity; poverty; population growth; and ecological degradation.[32] Creating such a government agency would underscore a "strict fact of life" in the contemporary world: national security cannot be sustainably achieved or even sensibly pursued without attention to global security.

Much political and legal advocacy for global security is carried out by non-governmental organizations (NGOs). NGOs are neither political parties nor government agencies. Sometimes called non-profit organizations, they operate within civil society and constitute one of the primary means that citizens employ to promote public goods such as health, environmental protection, social equity, peace, human rights, education, and cultural exchange. It is estimated that there are over 40,000 international NGOs currently operating, with local, regional, and national NGOs numbering in the tens of millions.

Historically, NGOs were crucial participants in many progressive movements, such as the effort to end slavery in the early 1800s, the women's suffrage movement of the early 1900s, and the disarmament and peace efforts organized later that century. In 1886, America's first popular conservation organization, the Audubon Society, was formed. Its mission was to protect game birds from overzealous hunters and plumage birds, such as egrets, from the then thriving millinery (hat) industry. The Plumage League, later renamed the Royal Society for the Protection of Birds (RSPB), was formed in the United Kingdom three years later. Like the Audubon Society, the vast majority of RSPB members were women outraged by the plight of young birds left to starve in the nest after their parents were shot for plumes that were fashionable in Victorian-era hats.

The Sierra Club, which focused on the protection of California's (and later the nation's) scenic and wild places, was founded in 1892. Across the Atlantic, in 1895, the National Trust for Places of Historic Interest or Natural Beauty was incorporated to educate citizens and help preserve the scenic and heritage sites of England, Wales, and Northern Ireland. More than a dozen National Trusts with conservation mandates have been founded in other countries since then. Between 1901 and 1960, an average of three conservation-oriented NGOs formed each year in the United States, with similar levels of growth in many European countries. That number would more than quintuple in the following decades, signaling the birth of the modern environmental movement.

Environmental and sustainability NGOs play crucial roles in protecting the natural world and promoting social benefits. They engage in scientific research, political advocacy, public education, and hands-on efforts to secure social welfare and environmental protection. Many of the larger environmental NGOs have memberships in the millions and annual budgets ranging over US$100 million. They are run by professional administrators and have departments of scientists, lobbyists, lawyers, public relations personnel, fund-raisers, and membership recruiters at their disposal. They organize citizens at the grassroots, lobby governmental officials, support political candidates for office, work in tandem with and in opposition to business corporations through partnerships and boycotts, employ the court system to seek legal redress both domestically and internationally, and frequently exploit the "court of public opinion" through extensive media campaigns. Other NGOs are small, local groups of concerned citizens who volunteer their time and energy to promote a particular public interest.

A prominent environmental NGO is the International Union for the Conservation of Nature and Natural Resources (IUCN). Founded in 1948 by both governmental actors and civil society organizations, the IUCN lays claim to being "the world's oldest and largest global environmental organization." Headquartered in Switzerland, the IUCN is affiliated with more than two hundred governmental and more than nine hundred non-governmental organizations. It is one of a score of NGOs claiming a formal association to the United Nations Educational, Scientific and Cultural Organization (UNESCO), which formed in 1946 with the purpose of promoting intellectual collaboration, education, social justice, and conservation. The IUCN was responsible for the formation of CITES, and publishes the most comprehensive and authoritative listing of the world's threatened and endangered species of plants and animals, known as the Red List. Other international NGOs with similar conservation mandates include the World Wide Fund for Nature (WWF), founded in 1961 as the World Wildlife Fund, and Conservation International, founded in 1987.

A more confrontational and controversial environmental NGO is Greenpeace. Greenpeace grew out of the efforts of a small group of citizens who opposed the testing of US nuclear weapons in the Aleutian Islands off Alaska in 1970. These peace activists also had green values, hence the name Greenpeace. Further opposition to France's nuclear tests in the South Pacific in the 1980s led to French secret agents sinking a Greenpeace ship docked in Auckland, New Zealand, causing the drowning of a Greenpeace photographer. The subsequent public uproar led to the resignation of the French defense minister and pushed Greenpeace into the international limelight, helping it become the largest member-based environmental organization in the world, with nearly three million supporters. Greenpeace's creative and very public campaigns against whaling, the fur seal trade, the ocean dumping of waste, and toxic pollution by large corporations have gained global support and no small share of notoriety.

Greenpeace inaugurated an era of environmental "direct action." While eschewing violence and the destruction of property, its activists frequently engage in civil disobedience. They have sailed into nuclear testing zones, put their rubber dinghies between whales and whaling ships armed with harpoons, and climbed smokestacks and skyscrapers to release huge banners condemning corporate polluters. Greenpeace has never shied away from taking the role of David against the Goliaths of multinational corporations and nation-states. Exploiting and often creating controversy as a means of bringing issues of concern to public attention, Greenpeace employs media savvy to achieve its purposes.

The NGO "community" has become a large and effective force for the promotion of social justice and environmental protection. All three of the United Nations global sustainability conferences – the 1992 Earth Summit, the 2002 World Summit on Sustainable Development, and the "Rio + 20" Earth Summit in 2012 – had parallel forums arranged for NGOs that were attended by thousands of representatives. More than nine thousand NGOs participated at the 2012 Rio conference, where NGO representatives also convened a parallel "People's Summit" a few miles away from where government delegations held official meetings. The World Social Forum is another NGO convention set up to parallel, and rival, an official meeting of corporate and governmental leaders. Under the motto, "Another World is Possible," the World Social Forum annually gains widespread representation from the NGO community to promote social justice, indigenous and human rights, and environmental conservation. Participants gather in various cities across the globe each year to advocate their causes, support fellow activists, and present an alternative vision to unsustainable economic policies and the growing corporate power they believe to be represented at the World Economic Forum, held in Davos, Switzerland.

As the activities of NGOs demonstrate, the politics of sustainability go well beyond the actions and agendas of governments. It concerns governance. Governance is the guiding or steering of behavior. NGOs do not govern as city councils, state legislatures, national assemblies, prime ministers, and presidents do. But they are powerful forces involved in governance at local, national, and global scales. They inform, educate, empower, and mobilize the public as citizens and consumers, and they engage in the architecture of choice. The Global Ecolabelling Network (GEN), for example, is an association of organizations that promotes environmental performance monitoring, certification, and labeling worldwide, allowing consumers to distinguish brands and products by their social and environmental impacts. Other organizations, such as Goodguide, rank products, as well as business corporations, based on their environmental and social impacts. An analogous service in the political realm is provided by the League of Conservation Voters, which publishes a National Environmental Scorecard each year, ranking the voting records of all members of the United States Congress. A "Dirty Dozen" list highlights those representatives with particularly bad environmental records. Making such information transparent empowers consumers and citizens.

Sustainability politics does not require more government. It does entail more governance and a vibrant civil society.[33] As Lester Brown observes, "saving civilization is not a spectator sport" and it cannot be accomplished by individuals engaged solely in "lifestyle changes."[34] Practicing sustainability entails governance carried out in full acknowledgment of our global interdependencies.

Aristotle defined the human being as a *zoon politikon*, a political animal. His point was that no individual is wholly self-reliant. To be human is to be interdependent, a member of a community who can fully develop only through regular and supportive interactions. Sustainability is grounded in the social, environmental, and economic interdependencies of the world's citizens, and these interdependencies necessarily require navigation through political means.

There is an old Southern African saying: *umuntu ngumuntu ngabantu*, or *ubuntu* for short. It means "I am what I am because of who we are," or more generally, "A person is a person because of people." *Ubuntu* is an acknowledgment of human interconnectedness. Archbishop Desmond Tutu observes that the *ubuntu* philosophy is captured in the belief that "I am a human because I belong. I participate. I share."[35] In short, we exhibit our humanity when we actively sustain the community that sustains us. No one can live a sustainable life on his or her own. Our active participation in families, communities, businesses, civic associations, and politics is required.

Sustainability demands political activity and legislation because bad politics and law are the cause, or at least the all-too-willing accomplices, of unsustainable economies and lifestyles. Not all advocates of sustainability endorse political and legal engagement. Bill Mollison observes:

> The tragic reality is that very few sustainable systems are designed or applied by those who hold power, and the reason for this is obvious and simple: to let people arrange their own food, energy, and shelter is to lose economic and political control over them. We should cease to look to power structures, hierarchical systems, or governments to help us, and devise ways to help ourselves.[36]

Politics often serves as a means to gain personal advantage at public expense; laws frequently serve to protect privilege. Given this reality, Mollison and many others place all their hopes for sustainability in local efforts that steer clear of political and legal entanglements. Mollison's endorsement of community self-reliance – of local governance rather than centralized government – is stirring. But he overstates the case. We cannot practice sustainability in the absence of collective action, and we hamstring ourselves severely if this collective action is restricted to the most local of communities. Sustainability cannot effectively be pursued, certainly not at national or global levels, without governments. It requires their use – and their transformation – to ensure greater transparency and citizen empowerment. The practice of sustainability weds the self-reliance of empowered citizens and communities with the skillful utilization of political and legal mechanisms.

Sustainability is both a political and legal challenge and an exercise in self-reliance. While self-reliance has obvious limits in a connected world, it is indispensable. Self-reliance prevents dynamic connections of interdependence from degrading into inequitable bonds of dependence. Sustainability rests on social equity, and the only means of ensuring that people are treated equitably is to empower them.

The political and legal challenge today requires citizens to take greater responsibility for – and foster the resilience of – their personal lives, their communities, their nations, and their common world. To this end, leadership must be bettered, effective policies crafted, good laws written, political and legal institutions held accountable or transformed, civil society enlivened, and empowering forms of governance developed.

Inquire and Explore

1. How can you engage in sustainability politics and practice sustainability leadership?
2. Which of the products that you purchase are green and healthy, and which ones are not?
3. What is your ecological footprint, and how can you shrink it?

* * * * * *

- Discover what you have the "right to know."
- Investigate the politics of climate change, and learn how you can make a difference.

 For web resources and more information on these topics, please visit conservationandcreativity.net

6
Sustaining Economies

An economy is a system of relationships focused on the production, distribution, exchange, and consumption of goods and services. Each of these activities exploits resources and, at least potentially, generates waste. In turn, each of these activities impacts the social system. The benefits of a strong economy are clear: economic relationships help meet basic needs, develop social networks, facilitate cultural development, spur technological innovation, and foster prosperity. Historically, economic enterprise has also greatly damaged the social and natural world. Today, this destructiveness occurs at a global scale. The extinction of countless species, the pollution of water, land, and air, the devastation of forests, the erosion and desertification of land, the depletion of sea life, the paving over of green space, and global climate change are all products of the way we run our economies. The manner by which we generate jobs and wealth, ironically, often undermines our ability to sustain life and health.

Most human enterprises have to be economically viable to persist. For a business to survive, sales revenues must cover expenses. More than that, profits must be made. Without profit, there is little incentive to run a business. That is the bottom line.

In a competitive economy, it is not easy for a business to be profitable. More than a third of new businesses fail within two years; and more than half of all businesses fail within four years.[1] These statistics vary some, but not much, depending upon the sector or industry (e.g., finance, manufacturing, leisure and hospitality, construction, education, and health). So, despite the best efforts of those starting up economic enterprises, most cannot maintain them profitably – if they ever were profitable – for more than a few dozen months. This is a sobering fact. And the challenges for business are compounded if profits cannot be gained at the expense of the environment and social justice. To keep a business alive today is to fight the odds.

When sustainability is discussed in the context of economy, the difficulties faced by business people struggling to make a profit in a competitive market are often ignored. That is a mistake. Sustainability advocates will not gain sufficient support from the political leadership and the business community until they concern themselves with the business of staying in business. Sustainability rests on the triple bottom line of "people, planet, and profit." To ignore the challenges faced by business is to neglect a mainstay of sustainability.

An increasing number of books and journal articles explain how profits can be maintained or increased without hurting people or harming the planet.[2] This scholarship provides useful references, solid facts, and often inspiring stories from the business world. And there are many encouraging signs. Increasing numbers of large businesses now have sustainability units, chief sustainability officers, and regularly employ sustainability consulting. Indeed, more than two thirds of business managers now say that sustainability is crucial to their competitive success in the marketplace.[3]

This chapter does not provide a how-to manual for transforming green into gold. Here, we take on a different task: investigating why creating a sustainable economy – one that delivers an equitable and environmentally benign production, distribution, exchange, and consumption of goods and services – may be the toughest and most important challenge humankind has ever faced. We start by directly confronting the near-universal belief that economic growth is required to sustain prosperous businesses and societies. The impacts of growth-based economic systems that rely on high levels of consumption and waste are then examined. An exploration of the developing field of ecological economics provides an alternative to business as usual. Placing these insights within a global setting, we conclude with an examination of international trade and the impact of the size of business ventures, from large multinational corporations to small, community-based enterprises.

Economic growth

Economic systems grounded in capitalism and free markets have expanded rapidly in the last two centuries, largely owing to four factors: the discovery and exploitation of previously untapped natural and human resources, including those of the Americas, Africa, and Asia; relatively cheap and abundant energy sources, chiefly gained from fossil fuels and most predominantly coal, oil, and more recently natural gas; technological development, particularly in mechanization, transportation, infrastructure, and communication which facilitated resource exploitation, production, and trade; and growth in human populations, which has steadily increased both the supplies of and demand for goods and services.

During this period, largely as a result of market economies utilizing cheap, abundant fossil fuels and producing more housing, food, clean water, and medical care, many forms of human welfare have improved. Today, on average, people live more than twice as long and enjoy over eight times more goods and services than their forebears of the 1800s.[4] Not surprisingly under these circumstances, the last two centuries also raised the human population sixfold. To vastly increase the goods and services available to a much larger population of much longer-lived individuals is an amazing achievement. We might celebrate our species for this historic economic growth, nutritional improvement, and medical progress. The only problem – which might dim the lights some on our celebration – is that we live on a finite planet.

Over the last two decades, most national economies have grown at over 4 percent a year and seldom fall below 3 percent a year. Many developing economies grew much faster, with China's economy often growing at nearly three times that rate. Much economic growth translates into increased *throughput*: the amount of natural resources being extracted from the earth, milled, refined or otherwise processed, manufactured into goods, distributed for consumption, and discarded after use. Throughput is the input of raw materials and energy combined with the output of goods produced and waste generated. It is a measure of the flows of raw material, energy, and waste materials into and out of the system. For the last few hundred years, we have grown our economies chiefly by increasing throughput. This process cannot indefinitely continue on a finite planet.[5] Physical and biological limits forbid it.

An old story makes this point well. An ancient wise man had done an important service for an Indian king. As a reward, the king offered him anything he wanted, within reason. The wise man, observing that the monarch had been playing chess, asked that a single grain of rice be placed on the board's corner square, and that the number of rice grains be doubled on each of the remaining 63 squares. This rice would be his sole reward.

The king was much amused by the request, and quite pleased, for this was a reward he thought most easy to supply. He set his court attendant to the task. The attendant made a brief trip to the kitchen and returned with a small sack of rice, placing 255 grains on the eight squares of the first row of the chessboard (1 +2 + 4 + 8 + 16 + 32 + 64 +128 = 255). The next few rows exhausted available supplies from the kitchen. The king ordered his attendants to the state's grain silos.

As the silos were being emptied, the alarmed king called in the court mathematician, and a quick calculation confirmed his worst fears. The 64th and last square of the chessboard would require billions of grains. Indeed, the total amount of rice on the chessboard would constitute a heap larger than Mount Everest and weigh well over 100 billion tons. Spread out, the 18,446,744,073,709,551,615 grains of rice the wise man was promised would cover the entire surface of India to the height of the king himself.

Clearly, anything that grows in a geometric progression, doubling in size on a regular rate, quickly exceeds the physical limits of a finite space. To figure out the approximate period of time in which something doubles in size, one simply divides 72 by its growth rate. So a country with an economic growth rate of 10 percent – a rate many countries managed or exceeded in the previous two decades – would double its GDP every 7.2 years.

Imagine, then, that the United States would be in what most of its politicians and economists consider the very enviable position of growing its economy at 10 percent a year. If its economic growth continued to be matched by an equal rise in its consumption of natural resources, then we can relatively easily calculate the amount of raw materials that would be consumed over any given period. In 2000, the United States consumed 6.5 billion metric tons of raw materials (including industrial minerals, fuels, natural aggregates, primary metals, wood, and other organics).[6] Let's start with this figure. If we double that level of consumption every 7.2 years, the country would eat through 6×10^{24} kg of raw materials in less than 300 years. That sounds like a lot, and it should. It equals the weight of the planet![7]

In advanced economies, growth in GDP does not translate into an equal growth in the consumption of raw materials, as much GDP may be derived from services and exchanges or increased efficiency. Throughput can also be reduced by dematerialization, the using of fewer material resources to achieve the same level of functionality. For example, email exchanges can reduce the need for paper and postage; electronic downloads of music can reduce the need for CDs; and new and better services can reduce the need for physical products, as occurs when the sharing, renting, or leasing of many occasional-use items, such as bicycles, cars, farm machinery, or power tools, reduces the need for their purchase and ownership. In turn, productivity gained by way of increased efficiency may create more wealth with fewer resources. But there is no evidence that increased efficiency alone will be able to maintain economic growth without also requiring steady increases in throughput.[8] And, notwithstanding creative efforts at dematerialization, the exploitation of most natural resources is steadily increasing.

Simple mathematics demonstrates that economic growth based on high levels of throughput cannot continue indefinitely without quickly surpassing biophysical limits. Growth-based economics is a short-term strategy for accelerated development of production systems and built environments. It is not a long-term solution for sustaining the earth as a livable home.

In terms of waste and pollution alone, the impact of economic systems based on high volumes of throughput is staggering. As little as 5 percent of the raw materials that are taken from the earth end up in the goods we purchase. The rest becomes extraction and manufacturing waste. In other words, for every pound of durable goods we buy, as much as 20 pounds of waste are generated. And the goods that we purchase often quickly become waste as well, ending up sooner rather than later in landfills or incinerators. In the United States, about a million pounds of waste is generated for every resident of the nation each year. That amounts to 20 times each person's body weight per day.[9] Notwithstanding recent gains in efficiency and recycling, modern economies continue to generate increasing amounts of waste.

This is no accident. Consumer economies are explicitly geared to raising throughput. Victor Lebow, in a 1959 edition of *The New York Journal of Retailing*, wrote: "Our enormously productive economy...demands that we make consumption our way of life, that we convert the buying and use of goods into rituals, that we seek our spiritual satisfaction, our ego satisfaction, in consumption....We need things consumed, burned up, worn out, replaced and discarded at an ever increasing rate."[10] Few politicians today would be so frank as to endorse Lebow's description of a consumer economy. But the vast majority of politicians still promise to do their utmost to foster economic growth. And this growth – touted as the solution to most of the problems that societies face – is still largely generated by increases in consumption and waste accompanied by the depletion of resources.

While growth in GDP accurately captures the rise in value of market transactions, it neglects the environmental and social impact of the increased production, exchange, and consumption of goods and services. GDP measures the busyness of an economy. But it does not count many of the things that really count. For example, environmental disasters may actually increase a nation's GDP, as the market transactions involved in cleaning up the mess produce a net gain in economic activity. Likewise, the generation of toxic pollutants may contribute to the GDP if it forces downstream victims to pay the costs of cleaning up degraded environments. In turn, a healthy population that exercises, eats right, and engages in preventative medicine may contribute less to a nation's GDP than a more sickly population chronically in need of pharmaceutical, therapeutic, and surgical interventions. And, of course, people who abide by the dictum "Reduce, Reuse, and Recycle" contribute less to their nation's GDP than those who feverishly purchase and frequently discard large amounts of non-recyclable consumer goods.

As we observed in chapter 2, growing GDP does not mean greater social equity. Indeed, it often translates into greater concentrations of wealth, with complementary declines in social welfare. In turn, growing economies that produce greater wealth often have negative environmental impacts. That is because wealth promotes overconsumption and waste. The United States is the world's wealthiest country. And it is one of the largest, if not the largest, per capita consumer of natural resources and producer of pollution and waste, including greenhouse gases. At a global level, the wealth of nations correlates very strongly to the size of their ecological footprints. The amount of land that is required to sustain a population is largely a product of its affluence (the A in the $I = PAT$ formula). Wealthier countries, measured in GDP, generally consume far more than their fair share of the planet's resources and produce most of its pollution and waste. Clearly, a growing GDP does not mean all is well.

How is it that the seamy underside of economic growth gets consistently ignored? What William Ophuls observed over two decades ago remains true today: economic growth is the "secular religion" that provides "a social goal, a basis for political solidarity, and a source of individual motivation." Growth is an "all-purpose 'political solvent'" within which the differences between otherwise antagonistic political parties, social classes, and ideologies get dissolved.[11]

In the same vein, ecological economist Herman Daly observed that our economic system is "hooked" on growth because it serves as a substitute for greater equality. We are "addicted to growth" because the belief that we can grow indefinitely allows us to postpone grappling with the challenge of fairly meeting everyone's needs today. Advocates of growth are effectively updating Marie Antoinette's unconvincing solution for dire poverty: "Let them eat growth. Better yet, let the poor hope to eat growth in the future."[12]

Daly is referring to the famous (and likely fabricated) story of the wife of Louis XVI of France. Queen Marie Antoinette infamously suggested that the throngs of starving peasants demanding bread outside her palace in Versailles should eat cake – because cake was in abundance inside the palace. The French Revolution cut short the queen's plan for distributing baked treats to hungry peasants. It also cut short her life with the help of a guillotine. The bright future expected from growth-based economies, Daly suggests, is as illusory as Marie Antoinette's vision of cake for all.

Growing economies promise more jobs, higher per capita income, greater wealth and hence higher revenues from taxes and increased governmentally sponsored public goods. A growing economy is likened to be a rising tide that lifts all boats and makes for smoother sailing for everyone. We might note that the oceans are indeed rising. And the reason is economic growth – at least that growth made possible by the burning of fossil fuels and destruction of forests. But these rising seas are swamping many coastlines, and there is more stormy weather ahead.

Peak consumption and peak waste

Faced with mathematical laws and biophysical limits, future economies will not be able to prop themselves up by way of ever-increasing consumption and throughput. As Richard Heinberg cogently argues, we have hit the "end of growth," and now must adapt to a "new economic reality."[13] The new reality entails sustaining societies within the carrying capacities of natural resource bases. But this does not mean that societies must stall the pursuit of welfare or the development of culture. Indeed, sustainability goes hand in hand with social progress and cultural creativity.

In his *Principles of Political Economy*, published in 1848, John Stuart Mill wrote:

> It is scarcely necessary to remark that a stationary condition of capital and population implies no stationary state of human

improvement. There would be as much scope as ever for all kinds of mental culture, and moral and social progress; as much room for improving the Art of Living and much more likelihood of its being improved when minds ceased to be engrossed by the art of getting on.[14]

Herman Daly, who developed the notion of throughput, expanded on Mill's idea of a stationary economy and population within an ever-developing society and culture. Daly proposed that our goal should be a "steady-state economy."

A steady-state economy promotes the development of sociocultural well-being without increasing the consumption of natural resources and emission of pollutants. To the extent that technology contributes to this form of economy – by meeting basic needs, enhancing the quality of life, and decreasing throughput – it, too, may be indefinitely developed. Technological innovation and sustainability can and should be complementary goals.

The first step toward creating a steady-state economy is meeting the basic needs of citizens without increasing throughput. This is the task of sustainable development. Sustainable development has been defined as "improving the quality of human life while living within the carrying capacity of supporting ecosystems."[15] Unlike economic growth, development can continue indefinitely since there are endless ways of improving the quality of human life without increasing throughput. Growth pertains to an increase in number or size. Development refers to amelioration: a making better. As the Earth Charter stipulates, once "basic needs have been met, human development is primarily about being more, not having more."[16]

The language of sustainable development is frequently coopted. Often when the term is deployed, what is pursued is business-as-usual industrialization. In such cases, sustainable development becomes something of a ruse: social acceptance and political currency are gained in developing countries for projects of export-oriented economic growth at the expense of meaningful gains in ecological conservation and social empowerment.[17] Sustainable development has been labeled a "great scam" whereby we help "some third world country 'develop' by selling the dignity of its people and their natural heritage for the benefit of others who lack for nothing."[18] The problem with promoting sustainable development, Donald Worster writes, is that it "will be 'development' that makes most of the decisions, and 'sustainable' will come trotting along, smiling and genial, unable to assert any firm leadership, complaining only about the pace of travel."[19] Carried out sustainably, economic development is the surest path to a steady-state economy. Without the "leadership" of sustainability, however, economic development tends to reinforce the inequities of the status quo and indefinitely defer environmental caretaking.

We have reached a stage in the planet's history when we can no longer pretend that economic development based on increased throughput can be sustained, even if it promises cake for all. Stocks of non-renewable natural resources (such as certain minerals and oil) are declining. And many renewable natural resources, such as fertile soil, forests, fisheries, and surface water, are being depleted far faster than their natural rates of replenishment. In turn, global levels of pollutants threaten what E. F. Schumacher called the "tolerance margins of nature," that is, the capacity of the biosphere to absorb toxins and waste. The tolerance margins of nature are quite resilient. If not unduly taxed, they can provide their benefits indefinitely. But when pushed too far their crucial services end.

A body of water may be able to dilute and absorb small amounts of pollution – say nitrogen and phosphorous run-off from nearby fertilized fields – without harming its productivity. But if the rate at which these pollutants are introduced exceeds tolerance margins for their absorption, algae that feed on the chemical nutrients will grow rapidly, depleting much of the dissolved oxygen. The body of water then becomes hypoxic, without sufficient oxygen to support many forms of life native to it. This occurs in many lakes, rivers, and estuaries subject to excessive run-off from agricultural lands. In the northern Gulf of Mexico, chemical nutrients flowing in from the Mississippi River create a 20,000 sq. km "dead zone" each year where sea life has been virtually extinguished. There are now 500 dead zones in the world's oceans, and they are growing in number.

Scientists have assessed the "planetary boundaries" that define the safe limits within which humanity can thrive within the earth's system.[20] These planetary boundaries are being approached or have already been exceeded in biodiversity loss, freshwater scarcity, land-use change and three forms of pollution: carbon dioxide pollution causing climate change and ocean acidification; various forms of chemical and particulate pollution of the land, water, air, and stratosphere; and nitrogen and phosphorus pollution. Pollution from the excessive use of nitrogen and phosphorus has already breached planetary boundaries, the scientists conclude, severely degrading fresh and coastal waters, accumulating in soils, and producing greenhouse gases, such as nitrous oxide. As a species we are generating more waste than the planet can safely absorb.

We are also consuming renewable resources faster than the planet can replenish them and non-renewable resources at rates that cannot be sustained. According to the International Energy Agency, the world reached its peak in conventional crude oil production in 2006. Peak oil has occurred because relatively few large sub-surface reserves are being discovered, and existing reserves are steadily being depleted. Indeed, we would need to discover new oil reserves every few years that rival in size the largest on the planet, such as Saudi Arabia's, just to keep pace with current rates of consumption.

Global demand for petroleum today is more than twice as much as it was half a century ago. Currently, the world consumes more than 80 million barrels of oil each day. Supplies of crude oil will be unable to match this output for long. That does not mean we will run out of oil any time soon. Shale oil (trapped between layers of rock that is difficult and more expensive to pump out of the ground), and oil sands or tar sands (sand or sandstone mixed with a very thick oil that takes a great deal of energy and water to refine), will increasingly be tapped as traditional sources of oil decline. But the steady supply of cheap oil that drove economic growth over the last half-century will not last forever.

Potentially even more problematic is the fact that we are living at a time of "peak water" where the supply of easily accessible freshwater on the planet cannot meet ever-rising demand. What oil was to the nineteenth and twentieth centuries, water will likely be to the twenty-first century: a source of ever-increasing demand and conflict. The insufficiency of freshwater supplies will hamstring economies and threaten survival.

Water is life. Without it, neither human beings, nor virtually any other species on earth, could survive. Freshwater is a quickly dwindling resource. There is no continent, with the possible exception of Antarctica, and virtually no country, that does not face regional water scarcity. In some developing countries, scarcity results from the lack of infrastructure to retrieve water from rivers or aquifers (e.g., pipes, pumps, and wells). Increasingly, however, water is physically scarce: available supplies, even with the best infrastructure, cannot meet the needs of human populations, economies, and ecosystems.

The World Health Organization estimates that one in three people across the globe currently suffers from water scarcity.[21] The United Nations predicts that by 2030, half of the world's people will live in areas facing acute water shortages, with the figure rising to two thirds of the global population by mid-century. In water-scarce areas, life will become harder, poorer, unhealthier, and shorter.

These hardships and dangers will occur largely owing to inadequate access to safe drinking water, which already affects almost a billion people. More than twice that number suffer from inadequate access to water for sanitation and waste disposal. This drastically increases the spread of dysentery, typhoid fever, and cholera, as well as other water-borne diseases. It is estimated that patients suffering from such infections already occupy up to half of the world's hospital beds.[22] In developing countries, four out of five illnesses are caused by sanitation issues related to lack of clean water. Growing water scarcity will also directly impact agricultural production; it will increase food shortages and hunger.

Food accounts for more than 70 percent of the world's water footprint – the total amount of water consumed per capita. Irrigation for agriculture uses up most of the available freshwater, especially in the developing world (up to 90 percent). So water scarcity translates directly into food scarcity. Everything we eat requires water to be grown, cleaned, and processed. Some foods are much more water hungry than others. Each kilogram of cereal (wheat, corn, or barley), tubers (potato, cassava, or yam), or other vegetables that is produced uses up to 2 cu. meters of water. Each kilogram of beef uses seven times as much, mostly because of the amount of grain that cattle consume. On average, to raise a cow to maturity requires 4,000 cu. meters of water. Depending how cattle are raised, their water footprint may be over a hundred times as high per unit as cereal production.[23] As more people include more meat in their diets, the global water footprint of food will rise significantly.

Like food, the durable goods we buy also have water footprints, as water is used in virtually every industrial process. Indeed, almost 60 percent of all water consumed in developed countries is used for industry. When we import durable goods, just as when we import agricultural goods, we are effectively importing water from other countries. A T-shirt requires 2,000 liters of water to produce; leather shoes require 8,000 liters.[24] About 20 percent of the more than 2,800 cubic-meter water footprint of US residents is imported. Of the near-1,400 cubic-meter water footprint of residents of Japan, almost 80 percent comes from beyond its borders.[25]

The average US resident uses, either directly or indirectly, about 1,800 gallons of freshwater a day. That is twice the global average.[26] In developed countries, people consume 30–50 times as much freshwater per capita as those in the developing world.[27] Growing populations, coupled with greater demand by agricultural and industrial sectors in developing countries, means that the consumption of water is increasing even if per capita use remains relatively low.

In past decades, as populations, agriculture, industry and the need for water grew, the general solution was to increase supplies to meet demand. This meant drilling more and deeper wells, and pumping more water out of rivers and lakes. But aquifers across the globe are now dangerously depleted, while rivers and lakes are increasingly tapped out.

With more than 97 percent of the world's water found in oceans, one might hope that technology will save us. Over 20,000 desalination plants are currently turning salt water into freshwater in more than 120 countries, including Israel, Saudi Arabia, United Arab Emirates, Singapore, Pakistan, Australia, and many US states, including Florida, Texas,

Arizona, and California. These are wonderful feats of engineering. And the number of such plants is likely to triple in the coming decades. No doubt desalination technology will also improve.

The problem is that desalination requires tremendous amounts of energy. Most of the energy used to desalinate (as well as pump and treat) water currently comes from fossil fuels, primarily coal. Coal power plants emit more carbon dioxide than all cars, trucks, buses, and trains combined. Adding the task of desalination to their workload would significantly heighten the emission of greenhouse gases.

So the increased need for desalinated water will only heighten the use of fossil fuels, which in turn will exacerbate climate change. Climate change will melt glaciers, and the absence of glaciers will bring drought to various regions – including already drought-prone countries such as India and China, whose rivers currently find their sources in the Himalayas. A vicious cycle is created. As the use of water becomes increasingly energy-intensive, fossil fuels burned to run desalination plants will worsen the climate change that threatens regional water supplies.

Water, unlike oil, is a renewable – and recyclable – resource. So, unlike oil, the amount of water on the planet is not declining. But the demand for freshwater, like that for oil, is steadily rising above available supplies. Aquifers, such as those in the Indian subcontinent or the Ogallala Aquifer beneath the High Plains in the United States, are being pumped out at rates many times higher than they are naturally recharged.[28] If current trends continue, it is only a matter of time before this "fossil groundwater" is effectively depleted. Many non-renewable and renewable natural resources, including water, crucial minerals, fertile soil, and tolerance margins, have peaked or are near peaking in this sense. Grappling with peak oil, peak water, peak minerals, peak soil, and peak pollution leads to a sobering conclusion: we are living at a time of peak consumption and peak waste. Politicians and pundits regularly propose the same solution for the social and economic problems their nations face: grow the economy. But attempting to solve the problem of peak consumption and peak waste by growing the economy is rather like trying to dig your way out of a hole.

Ecological economics

Actually, there is one way to dig your way out of a hole. Point your shovel sideways and begin the arduous process of crafting a stairway going up. That is precisely what sustainable development attempts to do. But to truly be sustainable, development must be firmly grounded in ecological economics. Ecological economics treats the economy as a subsystem of the environment. The assumption is that economies can only sustainably thrive if the biosphere remains healthy, fertile, and resilient.

A fundamental tenet of ecological economics is that natural capital should not be treated as if it were income. Income is money available to be spent. Capital is money (re)invested to provide the foundation or infrastructure for business. Once your capital is gone, you are out of business. Capital is like the money you put into a savings account to provide a monthly stipend derived from interest. To spend down the capital in the account means that there is no money left to earn interest. At that point, the stipend stops.

Natural capital is a stock of resources. In ecological economic terms, a forest is a stock of trees, other flora, and wildlife. Its timber may be used for lumber, energy production, and paper products. Trees also provide the service of carbon sequestration. Other forest flora may produce food, various services, and goods such as pharmaceuticals. Forest fauna may also provide foods and goods. When utilized in a sustainable manner, this stock of resources is not diminished. When exploited unsustainably, the stock becomes depleted, and the natural "income" that it would otherwise generate on a continuing basis ceases. At that point, with the forest gone, one is left with neither capital nor income.

Natural capital is composed of three things: the stock of the planet's non-renewable natural resources, such as fossil fuels and minerals; the stock of renewable natural resources, such as surface water, forests, and wildlife; and the capacity of the biosphere to absorb waste and pollution, its tolerance margins. Ever-growing economies that exhaust non-renewable resources and exploit renewable resources and tolerance margins faster than their rates of repair and replenishment create ecological debts.

Ecological economics aims to create thriving economies that do not deplete the natural capital upon which they depend. As we saw in chapter 3, the economic benefits that humans gain from nature's services are considerable. Robert Costanza, who spearheaded the effort to calculate the monetary value of these services, was one of the founders of the field of ecological economics. Another key figure was Nicholas Georgescu-Roegen, a Romanian economist who mentored Herman Daly after immigrating to the United States.

In his book, *The Entropy Law and the Economic Process* (1971), Georgescu-Roegen argued that the second law of thermodynamics, also known as the law of entropy, should be applied to economic affairs. The law of entropy states that in a closed physical system, energy tends to disperse and become increasingly unavailable for use. So, for instance, the energy in a cup of boiling water that is poured into a large pot of cold water will disperse throughout the pot and will no longer be available for use. It cannot be exploited to make a good cup of tea. Likewise, the concentrated energy of fossil fuels, once burned, is forever lost.

Georgescu-Roegen's work on entropy has convinced some ecological economists that man-made capital cannot replace natural capital. Human (technological) activity can never fully substitute for the concentrated forms of energy available from natural resources. Traditional, neoclassical economists tend to believe that technological innovations will allow economies to grow indefinitely by delivering replacements for materials and services that were once provided by nature. Ecological economists influenced by Georgescu-Roegen maintain that the stock of non-renewable and renewable natural resources, including the tolerance margins of nature and her other services, are irreplaceable. Once these concentrated forms of energy are wholly depleted, any attempt to create (technological) replacements will require a greater investment of energy than the system has available.

Can technology produce sufficient forms of concentrated energy, material resources, and services to replace depleted natural capital? Some argue that it is possible, at least in theory: we simply have to learn how to harvest solar energy efficiently, or perhaps develop atomic fusion. Others maintain that the law of entropy makes it physically impossible. In any case, we do know that a world depleted of natural capital would be fundamentally different from the world we use, love, and enjoy today. It is also clear, as we saw in chapter 4, that the technology developed to replace natural capital will produce side effects. And if history remains the best predictor of what the future will look like, these unintended consequences will present us with dire problems.

When technology causes unforeseen harm, we call it an unintended consequence. When a business enterprise designed to deliver some good produces harm, we call it an externality. An externality is a risk or harm generated as the by-product of economic activity. Businesses often externalize the social and environmental costs of production in order to maximize their profits. Externalities are not really unintended consequences in that their harms are typically foreseen. For instance, a business that makes widgets may generate air or water pollution from its factory. If this pollution is sent out of its smokestack into open air or through its effluent pipes into a nearby river, rather than being fully treated at the site, then the cost of remediating this pollution is being externalized to the surrounding community. Either the community must pay money to have its air or water cleaned, or it must pay the (health) costs of living with polluted air or water.

When businesses externalize costs, they are effectively fobbing off expenses on taxpayers and other stakeholders. They receive the financial benefits of exploiting natural resources while the social and ecological costs get distributed across society, or left to future generations. A United Nations report estimates that as much as a third of all corporate profits are generated by externalizing environmental costs.[29] This externalization of harms has been dubbed the "tragedy of the market." It is the hidden cost that "markets, in the service of private property, impose upon the commons."[30]

To avoid this tragedy, ecological economists maintain that the risks and costs of doing business should be internalized. They endorse the "polluter pays" and "depleter pays" principles. Those businesses responsible for pollution or natural resource depletion should absorb the costs of cleaning up the pollution or offsetting the depletion and pass these internalized costs on to consumers in higher prices for their goods and services. This is known as full-cost pricing.

Extended producer-responsibility programs provide an example of such principles in action. Here, manufacturers and distributors of consumer goods remain responsible for the waste streams they create. A disposal or recycling fee is automatically added to the price of the goods. The fee is used to offset the costs of recycling used products and trash disposal. Many states or provinces and some nations have put extended producer-responsibility programs in place, particularly for electronic goods.

Green taxes, such as carbon taxes, are another means to achieve a similar end. Here, a government levies a tax on goods and services that have environmentally or socially pernicious impacts. Tax revenues may then be wholly or partially devoted to the remediation of the environmental or social harm. A carbon tax would not have to result in higher overall levels of taxation. At present, most revenues for the US federal government are gained from taxing personal income, corporate profits, investment income, and labor (payroll taxes). A decrease in one or more of these taxes could offset some or all of a carbon tax. In this way, carbon emissions would become costlier, and hence would be reduced owing to market forces, without adding to the overall burden of taxpayers.

A carbon tax is really just a way to ensure the full-cost pricing of goods and services that produce greenhouse gases. If the full cost of remediating climate change were assessed, prices for carbon-producing products and services would be significantly higher. It is estimated that the price of gasoline and diesel in the United States would be three to four times higher, for example, if all of the externalities of gas and diesel-powered transportation were effectively internalized. That is to say, if what you paid at the pump actually reflected the cost of remediating pollution, including greenhouse gas emissions, as well as all the local, state, and federal expenditures devoted to building and maintaining highways, paying for police, ambulance, and medical services associated with highway traffic and accidents, gas and diesel prices would triple or quadruple. Drivers of automobiles and trucks are effectively receiving massive subsidies from taxpayers to defray the full cost of using their vehicles.

In his groundbreaking book, *The Ecology of Commerce*, Paul Hawken asserts that "Without doubt, the single most damaging aspect of the present economic system is that the expense of destroying the earth is largely absent from the prices set in the marketplace. A vital and key piece of information is therefore missing in all levels of the economy."[31] Ecological economics ensures that this information is present and accounted for. It makes the costs of doing business transparent. Full-cost pricing, also known as full-cost accounting, prevents the market from lying. It also encourages the efficient use of resources, as there would be no effective subsidies to stimulate wasteful practices. Many hundreds of billions of dollars of environmentally destructive subsidies are currently being paid out by taxpayers worldwide. The International Energy Agency estimates that fossil-fuel consumption subsidies were well over US$500 billion in 2013, over four times the value of subsidies going to renewable energy, and more than four times the amount invested in improving energy efficiency.[32]

With full-cost pricing in place, goods and services that do not generate high levels of throughput and do not externalize costs and harms would find themselves more competitively priced. Rather than have government regulate businesses, full-cost pricing employs market mechanisms to foster "design for environment" strategies and spur technological innovation. Businesses, guided by nothing but the profit motive and the need to stay competitive, would adapt their technologies and use of materials to minimize social and environmental impacts. Full-cost pricing would stimulate green product design and green technology, as those designs and innovations that most efficiently deliver goods and services without externalizing risks and costs would gain a cost advantage in the marketplace. That is the business of sustainability.

Determining full-cost prices entails complex analyses and calculations. It requires a "life-cycle assessment" or "life -cycle analysis" of the environmental and social impacts of goods and services. Life-cycle assessment gathers and analyzes data on the social and environmental impacts of products and services "cradle to grave." It starts with the extraction of raw materials used to make goods and ends with the disposal of trash when used goods are thrown away.

In turn, life-cycle assessment measures and analyzes the impact of manufacturing processes and product distribution, including the toxicity of any chemicals or compounds employed, as well as the impact of pollution. It also accounts for the depletion of natural resources. Given that there are an estimated ten billion distinct products currently in the marketplace, with millions added every week, determining full-cost prices would be impossible if one had to conduct a life-cycle assessment of each product from scratch. Still, it is feasible to determine the full-cost price of the energy and basic raw materials or feedstocks that are used in producing most goods and services. If life-cycle assessment is conducted to full-cost price energy and feedstocks, then the goods and services employing these resources will closely approximate their full-cost price.

Life-cycle assessment and full-cost pricing are demanding and admittedly imprecise endeavors. It is impossible to measure all of the environmental and social impacts that a product or service may have, especially if one attempts to incorporate more intangible values that affect the quality of life. How do you put a price on aesthetic harm done by factory lights that prevent stargazing at night, or billboards that block scenic vistas on highways? And how do you put a price on the welfare of future generations? The danger is that such "soft variables" will get undervalued in any calculation. Still, that which is not measured will likely remain an externality. So the alternative to good faith efforts in life-cycle assessments and full-cost pricing is a world in which a lack of transparency allows economies to undermine the welfare they are meant to promote.

Businesses guided by ecological economics within steady-state economies would not be guaranteed success. As in any competitive market system, many businesses would fail. Yet failure can be a force for adaptive change. Ecological economies, no less than unsustainable ones, would depend upon the energy and innovation of entrepreneurs, people who take risks (with their own money) to turn new ideas into goods and services. As the economist Joseph Schumpeter observed about market systems in his book *Capitalism, Socialism and Democracy* (1942), entrepreneurs are sources of "creative destruction." They create new products and businesses processes that outcompete current models, producing a dynamic marketplace.

An ecological economy, like an ecological community of plants and animals, is an adaptive system. Economic development through adaptive innovation is beneficial, as long as these innovations do not gain market share by effectively hiding associated costs and harms. An ecological economy stimulates adaptation by eliminating, as much as possible, the externalization of social and environmental costs and harms within a competitive marketplace. Its market is both free and transparent.

Free trade and economies of scale

In low-income, developing societies, economic growth often produces increased environmental degradation. Here, new industries and businesses gaining a foothold are allowed to increase their profits by externalizing costs. As a country's economy further develops, certain forms of environmental degradation tend to decrease. Residents whose basic needs are mostly met come to demand increased protection from environmental harms. A much-studied economic theory, known as the Environmental Kuznets Curve (EKC), first established this relationship.[33]

Simon Kuznets, an economist, hypothesized that economic inequality increases over time in a developing nation, then peaks and begins to decline once a certain per capita income is achieved. An inverted "U" curve, the Kuznets Curve, illustrates this relationship graphically, as economic inequality rises up the inverted U to a tipping point, and then quickly drops. The hypothesis has been much disputed by economists who study the relationship between economic growth and equality. The rise in inequality in the United States in recent decades certainly puts the Kuznets Curve into question.

Better established is the inverted U-shaped Environmental Kuznets Curve that portrays the relationship between environmental degradation and per capita income. The EKC graphically presents pollution and other forms of environmental degradation rising over time as an economy grows and per capita income increases. Eventually environmental degradation peaks. It then begins a steep descent as the economy and incomes continue to grow.

The EKC hypothesis is subject to ongoing debate. But data generally confirm the relationship, at least in particular cases. An EKC exists for specific forms of air and water pollution, such as that caused by sulfur dioxide, nitrogen oxides, lead, DDT, particulate matter (like soot), and sewage. These forms of pollution diminish in countries with high per capita income. Data do not confirm the EKC relationship for other forms of environmental pollution and degradation, such as the loss of biodiversity, natural resource depletion, municipal waste production, or greenhouse gas emissions. And ecological footprints as a whole do not decline with rising incomes in growing economies.

Studies of the EKC offer little support for the conclusion that economic growth in itself is an environmental boon. Rather, rising incomes and accompanying political empowerment in developed nations generate policies to limit specific types of pollutants, namely those that create highly visible, localized problems – pollution that has immediate and palpable health effects on residents. But growing economies generally do not remediate, and often exacerbate, forms of environmental degradation whose consequences are more broadly distributed across generations and the global commons. Nations with developed economies and high incomes, for instance, are the world's highest per capita contributors of greenhouse gases.

In turn, highly developed economies often do not so much diminish as displace environmental degradation. Within a globalized marketplace, regulations in developed nations that reduce localized forms of environmental harm may effectively shift some of this harm to developing societies. Studies suggest, for instance, that deforestation decreases in highly developed economies but per capita use of forest products actually increases. In part, this increased appetite for forest products is satisfied through the import of wood, paper, and other forest products from developing countries, where unsustainable logging – including that of old-growth forests – is rampant. In such cases, deforestation is effectively being exported. Likewise, highly polluting industries may relocate to developing nations, where regulations for environmental protection are weak and the costs of production remain low. These same industries then sell their goods in the developed world. Here again, pollution is effectively being exported from high-income to low-income countries.

This brings us to the issue of trade. Economists developed the theory of comparative advantage to describe how free trade stimulates countries to specialize in the delivery of goods and services that they are particularly efficient at producing. First articulated by the economist David Ricardo in the early nineteenth century, the theory states that trade can benefit all involved if countries engage in the production of goods for which they are best suited. Tropical nations grow bananas and mangoes more efficiently than would be possible in northern countries, where heated greenhouses would be required. And northern countries, in turn, might be better at producing wool and pine timber.

The theory of comparative advantage demonstrates that trade between two countries may still be mutually beneficial even if one country proves more efficient (that is, has an absolute advantage) at producing everything. As long as each country is better at producing some things rather than others, both countries still benefit from trade if each focuses on what it does most cost-effectively.

The theory of comparative advantage would suggest that countries with the fewest and least stringent laws for environmental protection are the best places for polluting industries to locate. That is because the absence of environmental laws allows industries to externalize costs and produce their goods more cheaply. Free trade regimes generally forbid countries from levying tariffs or surtaxes on imported goods based on the pollution their extraction and manufacturing generates. As a result, free trade may displace rather than diminish environmental and social harms.

Absent regulations to protect the environment and social welfare, or full-cost accounting which produces the same protections through market mechanisms, free trade may prompt a "race to the bottom." Here, countries compete against each other to attract businesses by decreasing or eliminating environmental, labor, and health regulations that would eat into corporate profit margins. The predicted result is an overall lowering of regulatory standards and an overall increase in environmental degradation and social risk.

There is considerable debate as to whether, or how frequently, races to the bottom actually occur. Countries with strong regulations often remain very attractive to business for sundry reasons. Corporations often benefit from stable regulatory mechanisms, which make their operational and competitive framework more predictable. In turn, social and political cultures in key states may block the gutting of regulatory norms, effectively preventing a downward spiral.[34] The "race to the bottom" is not a fact of life in an era of global free trade. But it remains a lurking threat.

Economic globalization has many dangers and downsides. The increasingly meshed economic world it creates leaves people dependent on distant businesses for their basic needs and livelihoods. It can deprive local communities of the autonomy to shape their own future. And as we have just seen, in the absence of regulations within zones of free trade, it can weaken the ability of participating nations to protect the health and welfare of their citizens and environments.

For all these reasons, advocates of sustainability often promote the localization of economies. They encourage people to "buy locally" and "eat locally," supporting small businesses and regional agriculture and manufacturing. Efforts to promote local economies face tremendous obstacles from globalization and growing corporate power. Today, multinational businesses boast revenues that surpass the GDPs of many nation-states. Indeed, of the largest hundred economies in the world, over half belong to corporations rather than countries. Empowering the local is difficult given the economies of scale that generate much of this corporate wealth and power.

A producer's average cost per unit typically falls as the scale of output increases. This occurs owing to the capacity of large producers to buy materials and machinery cheaply in bulk, reduce production costs owing to greater access to advanced technology, benefit from efficiencies gained from the specialization of workforces and management, reduce advertising and marketing costs per unit sold owing to access to larger ranges of consumer outlets, and secure cheaper capital (lower interest on bank loans). Such economies of scale generally favor large corporations. In some sectors, such as manufacturing, the productivity of large firms (of more than 250 workers) is almost double that of small firms (of fewer than 20 workers).[35] The theory of comparative advantage appears to favor big business, as economies of scale allow for greater efficiencies in production and therefore lower costs.

To the extent that more resources are required and more waste generated when small, local businesses produce goods and services, sustainability is not well served by localizing. Economies of scale should be considered. It would not be a good idea, for instance, for every community to build and run its own steel smelter or automobile factory. At the same time, we should not let the higher costs of some locally produced goods and services persuade us that highly productive large-scale ventures are inherently more sustainable. Again, life-cycle assessments and full-cost accounting are required to determine ecological footprints and sustainability impacts. The actual cost of goods and services in the marketplace absent this assessment and accounting is not a reliable measure of true costs, as it does not reflect the resource depletion, environmental degradation, and social harms caused by its corporate producer.

In 1973, the British economist E. F. Schumacher published the prizewinning and highly influential book *Small is Beautiful: A Study of Economics as if People Mattered*. Schumacher was one of the first people to argue that modern economies were unsustainable because they treated natural capital as if it were income. He maintained that the goal of economic life was not to maximize output, income, or spending, but to achieve the greatest amount of well-being with the minimal amount of consumption. Opposed to the notions that "growth is good" and "bigger is better," Schumacher argued not for economies of scale but for appropriately scaled economies. Enabled by appropriate technologies, appropriately scaled economies can nourish local communities while safeguarding the environment.

The question of economic scale remains crucial today. Efficiencies in production and low costs to consumers will always remain vital concerns. But very large economic enterprises can deprive individuals and communities of the autonomy that is crucial to cultivating a sense of ecological and social responsibility. And in the absence of full-cost accounting, the sticker prices of their products rarely tell the whole truth.

Small can be beautiful. But big is not necessarily bad. Big can be efficient. And it can be creative. Large cities are places of great innovation, owing to their vast cultural, technological, and educational resources and the diversity of interactions they afford their residents. Sustainability entails acknowledging the demands of efficiency and the benefits of economies of scale without falling into the trap of thinking big is always better.

In turn, bonds of trust and care that make for resilient communities need not be weakened to appreciate our global opportunities and obligations. The practice of sustainability requires a glocal perspective. Local empowerment and responsibility may be strengthened without sacrificing the benefits of global connections, neglecting the salience of global problems, or denying the need for global solutions.

We cannot continue to run our economies in the coming decades in the same way that we have been producing, exchanging, and consuming goods and services for the past few centuries. To make such a statement is not to issue a moral imperative. It is simply to confront reality. Business as usual is not an option.

A shrinking, increasingly interdependent world can no longer tolerate exorbitant profits gained by externalized costs. Planetary stocks are depleted, and atmospheric sinks are full. Economic life, of necessity, will be radically different in the future. The opportunity before us is to make it both different and better.

Inquire and Explore

1. Is capitalism on a collision course with climate stability and ecological health?
2. How much do environmental externalities cost you, and how does life-cycle assessment work?

* * * * * *

- Can putting a price on carbon lower emissions without damaging the economy?
- Discover how to make people and planet matter while still making a profit.
- Learn about the art and craft of sustainable business.

 For web resources and more information on these topics, please visit conservationandcreativity.net.

7
Culture and Change

Culture is the world that humans make, enjoy, and improve through science and knowledge, ethics and politics, economy and technology, customs and diet, arts and recreation, religion and spirituality. The word stems from the Latin, *cultura*, which means to cultivate. Culture refers to the ways we cultivate a uniquely human place within the natural world. Among other things, culture establishes sets of practices and traditions that orient us to three fundamental ideals: truth, beauty, and goodness.

Practices of science and humanistic education apply themselves to the question of the true. They develop knowledge and understanding, guiding our natural curiosity, our inclination for discovery, and our propensity to solve problems. Artistic traditions apply themselves to the appreciation and creation of beauty. As human beings are naturally attracted to beautiful things, aesthetics becomes a crucial component of the cultural shaping of behavior. Moral and religious traditions apply themselves to the question of the good. They address what it is right and proper for us to think, say, and do. They tell us how we ought to act in light of our duties and obligations as members of particular communities, as members of the human race, and as spiritual beings.

In this chapter, we explore the development of a culture of sustainability. A culture of sustainability relies heavily on science to guide its development. But to live sustainably, we also need to develop a sense of place and purpose. At a cultural level, the how of sustainability is also a why. That is to say, the culture of sustainability situates us within sets of practices that foster a meaningful life. Accordingly, this chapter examines the ethical and spiritual foundations of a culture of sustainability. In turn, we look at the crucial impact of aesthetic sensibilities and education. The chapter concludes by exploring the role of creativity, examining how a resilient culture can maintain core values and relationships while adapting to a changing world.

Science and stories

Sustainability is not pie in the sky. It is a level-headed practice. One might say that sustainability transforms common sense into business acumen by bringing transparency and foresight to our economic activities. It forces the market to tell the ecological and social truth. We may ignore such truths in the short term. But to improve our long-term welfare, we must heed these truths and participate in their ongoing discovery. Ignoring them is like ignoring the law of gravity: it may produce a temporary sense of freedom to jump off a cliff, but the eventual impact will be devastating.

Ecological, economic, and social truths are best discovered and confirmed by thoughtful inquiry that incorporates diverse perspectives. The investigation of the manifold relationships that constitute our ecological, economic, and social lives is crucial to their understanding and improvement. Humanistic explorations of ethics and other aspects of human culture are crucial. The practice of sustainability is also wedded to rigorous inquiry by way of the natural and social sciences. A few examples of how difficult it is to live sustainably in the absence of such science may be helpful.

Let's say you are concerned with climate change and reducing your carbon emissions, and you are trying to figure out whether it would be better to drink a glass of milk or have a cup of hot tea. Life-cycle assessment is in order, and some rather sophisticated science is required. Not heating up the kettle for tea will certainly reduce fossil fuel use, keeping about 25 grams of carbon dioxide from entering the atmosphere. But you also have to take into account the way tea is grown, harvested, dried, and packaged, and how milk is produced, pasteurized, and refrigerated. Then you need to figure in the fact that the milk-producing cows belch prodigious amounts of methane, a potent greenhouse gas. Indeed, cows and sheep (which belch even more methane per pound than cows) generate about 10 percent of global greenhouse gas emissions worldwide. Some calculations soon produce the data you need. Drinking a glass of milk puts the equivalent of about 300 grams of carbon dioxide into the atmosphere. Drinking milk instead of tea is 12 times worse for our planet's climate.[1]

Now consider a dilemma faced by a locavore who is concerned with climate change. If you are committed to eating only local produce but still want to enjoy salads in winter, your only alternative is to consume produce grown in local hothouses if you live where the winters are cold. Yet studies demonstrate that local hothouse fruits and vegetables typically have a larger carbon footprint than imported produce grown in overseas fields, as ocean shipping adds only a twentieth of the greenhouse gas emissions involved in the production and distribution of most produce. Indeed, the transportation of produce from farm to marketplace contributes on average only 4 percent of the total greenhouse gas emissions related to food production.[2] With this in mind, locavores should figure in the impact of their own means of transportation to and from the vegetable stand. Driving a car to the farmers' market to pick up a few items generates a larger carbon footprint than walking or biking to the neighborhood supermarket to buy imported produce. In turn, consuming milk and meat (from methane-producing ruminants like cows and sheep) one day less each week would likely reduce your carbon footprint more than buying all locally produced food.

Most produce cannot be grown year round in a single locale, owing to specific climate needs. Water and soil requirements also matter. If Mexico were to grow all the wheat, corn, and sorghum that it currently imports from the United States, for example, this hot and generally dry country would have to find another 9 billion cubic meters of water for its agricultural fields.[3] Whether living in Mexico or elsewhere, locavores who want a varied diet but are also concerned about the efficient use of precious natural resources must do some refined calculations and comparisons.

Of course, the sustainability of our food depends on more than climate change and natural resource use. There are other environmental issues at stake in agricultural production, such as land and water pollution and the impact on biodiversity. In turn, there are social and economic concerns, such as agricultural worker wages and welfare. And that is to say nothing about the sustainability of the rest of our daily activities when we are not eating.

To the extent that we live, work, and consume in a world defined by complex, interdependent relationships, sustainability requires rigorous science. Not infrequently, this science will prompt us to change our beliefs and behavior. After all, that is why we do science: to learn. So we should welcome research and analysis – the best science of the day – when it contradicts our beliefs and forces us to revise assumptions, practices, and policies. Sustainability is an adaptive art, in large part because it is wedded to science.

But science is not enough. Notwithstanding its need for rigorous data gathering and analysis, the practice of sustainability entails moral commitment. It is grounded in ethical vision. And our moral commitments and ethical visions cannot be generated by scientific inquiry.

Science is crucial to the discovery of how we may consistently and effectively exercise our moral convictions. It helps us ensure that narrow self-interest does not masquerade as morality. It can protect us from hypocrisy and self-delusion. But science cannot instill the heart-felt dispositions that bring us to care for others and for the world around us. These dispositions arise from our immersion in families, friendships, and communities, webs of relationships both social and ecological. We discover what is right and proper to do by participating in networks of obligation and mutual aid. Effectively, our moral commitments and ethical visions develop over time as we come to appreciate our roles within the communities that sustain us.

How do we determine these roles? Moral philosopher Alasdair MacIntyre observed that "I can only answer the question 'What am I to do?' if I can answer the prior question 'Of what story or stories do I find myself a part?' "[4] To ask "What ought I do?" is to pose a question about duties and obligations. MacIntyre believes that our moral commitments – what we deem right and proper to do – are not determined by science, logic, or reason. Rather, these sensibilities grow within us as we discover our roles and find our place within particular relationships, communities, traditions, and sets of practices.

Our cultural lives – the integrated patterns of attitudes, beliefs, values, practices, relationships, and institutions that structure our existence and give it meaning – are essentially narrative in form. That is to say, we develop a sense of self, a sense of community, and the moral sensibilities that guide actions, by inhabiting living stories. Their plots are our daily practices. We learn our roles as the protagonists of stories.

To know what stories we partake of is to make our lives meaningful. Meaning is a complex blend of emotional attachments, cognitive beliefs, rational assessments, and imaginative projections. These attachments, beliefs, assessments, and projections get melded into a narrative framework that makes sense of our lives.

Thomas Berry, an "eco-theologian" and Catholic priest stated:

> We are in trouble just now because we do not have a good story. We are in between stories. The old story, the account of how the world came to be and how we fit into it, is no longer effective. Yet we have not learned the new story. ... We need a story that will educate us, a story that will heal, guide, and discipline us.[5]

Our cultural practices place us in narratives that provide a sense of direction and make our lives meaningful. Whether culture takes the form of intellectual, artistic, or ethical pursuits, it embeds us in networks of relationships. We navigate these relationships by understanding ourselves as protagonists of stories in progress.

Relatively few people today live within traditional cultures where belief systems, skills, and practices get handed down largely unchanged from generation to generation. Our role as cultural beings, therefore, is not simply to adopt a given narrative. The scripts of the narratives we inhabit are things we adapt. Likewise, the story of sustainability is open-ended. We are in the midst of crafting it. How the story of sustainability develops largely depends on the relationships we form within the local and global communities that sustain us.

Ethics and religion

We live ethically to the extent that we fulfill our duties and obligations and cultivate relationships of reciprocation and care. These responsibilities and relationships develop in response to emotional bonds established with those who share our world. They also develop as a product of our reasoned understanding of the roles and requirements we bear as members of particular communities. The community in question might be a family or circle of friends or colleagues, a particular neighborhood, village, or nation, the entire human species, or even the "community of life" as a whole. Aldo Leopold wrote that "All ethics so far evolved rest upon a single premise: that the individual is a member of a community of interdependent parts."[6] According to Leopold, an action is ethically right when it preserves the integrity, stability, and beauty of one's community. For Leopold, one's community extends beyond fellow human beings. He spoke of a "land ethic" that "enlarges the boundaries of the community to include soils, waters, plants, and animals, or collectively: the land."[7]

Leopold's description of the land community drives home an important point. People do not live in a community but in multiple, intersecting, or nested communities. The community of family is nested within the larger community of neighborhood, village, or town, and intersects with communities of friends and colleagues. These communities, in turn, are nested within national and global communities, and within the planetary community of life. While moral norms have varied dramatically in history and across cultures, it is fair to say that every moral system concerns the responsibilities that individuals have to sustain the communities that sustain them. What changes over time and space are the constituencies of these communities, the extent of their nesting, and the perceived requirements of their maintenance and transformation.

In an age of global interdependence, the communities that sustain us are multiple and expansive. Ecologically, our lives are sustained by biological, geological, and atmospheric relationships that span the globe. Economically and technologically, our lives are sustained by worldwide networks of commerce, trade, information, and communication. Politically, legally, and culturally, our lives are sustained by conventions, protocols, agreements, laws, alliances, organizations, institutions, and social networks that circle the planet. Martin Luther King, Jr, in his 1963 "Letter from a Birmingham Jail," wrote that "Injustice anywhere is a threat to justice everywhere. We are caught in an inescapable network of mutuality, tied in a single garment of destiny. Whatever affects one directly, affects all indirectly."[8] King's prescient pronouncement well captures the scope of an ethic of sustainability. The networks of mutuality that we inhabit today are glocal.

Attempts to formalize the complex and shifting relationships of mutuality that comprise ethical life are bound to fall short. There is no way fully to capture by way of principles or rules the vast, intricate, diverse, and shifting relationships and responsibilities that comprise our lives. But many attempts have been made, and they can be instructive and inspiring.

Perhaps the oldest and most widely accepted ethical principle is the so-called Golden Rule. Versions of the Golden Rule can be found within virtually every religion and cultural tradition. Two and a half millennia ago, Confucius maintained a deceptively simple principle by which to conduct one's life: "Do not impose on others what you yourself do not desire."[9] Similar statements are found within the Hindu, Buddhist, and Jewish traditions, among others. The Christian version is "Always treat others as you would like them to treat you."[10] It is a succinct, beautiful moral principle. But its simplicity hides a complex reality.

We should not presume that the needs and wants of others are identical to our own. Presenting a gift of alcohol to a devout Muslim friend, a baked ham to a Jewish colleague, a beef pot pie to a vegetarian boss, or a box of candy to a diabetic patient would be an inappropriate form of generosity. Clearly, we should not treat others exactly as we might want to be treated, that is, according to our own idiosyncratic tastes or context-specific needs. At the same time, it is fair to assume in the absence of direct evidence to the contrary that everyone wants his or her basic needs satisfied – physical health and security, clean water and ample nutrition, decent housing, education, economic opportunity, political empowerment, and environmental health and beauty.

A school of thought called utilitarianism holds that people want to maximize pleasure and minimize pain in their lives. The more pleasure and the less pain, the happier one's life. Utilitarians believe that to act morally is to promote the greatest happiness of the greatest number of people. Faced with a choice, the ethical person considers the consequences for all involved, and then acts to maximize pleasures and minimize pains over the long term for the majority of people.

Achieving the greatest good for the greatest number over the long term might constitute a passing definition of an ethic of sustainability, assuming that one assesses the direct and indirect environmental, economic, and social consequences of one's actions. The Australian moral philosopher Peter Singer has argued that a utilitarian framework should not be limited to human beings. It should be extended to all sentient life. Moral action, for Singer, contributes to the maximized well-being of any and all forms of life that can feel pleasure and pain.

Utilitarianism presents some quandaries. It is extraordinarily difficult to calculate with any certainty what actions will produce the greatest good for the greatest number over the long term. Human beings enter the world by means of a painful birth, suffer a good deal throughout their lives, and typically experience painful deaths. A utilitarian might conclude that minimizing pain is best achieved by anaesthetizing people, and drastically reducing their numbers. On the other hand, maximizing pleasure might be best achieved by increasing the number of people and other animals capable of experiencing pleasure, even to the point of massive overpopulation.

To complicate matters further, pleasure is often the product of painful efforts, as when hard labor produces pleasing results. Indeed, some forms of pain are themselves pleasurable: consider athletic exercises that push the body to its painful limits but produce a wonderful sense of exhilaration. And some experiences can be simultaneously pleasurable and painful: think of overeating or otherwise indulging to the point of feeling physically ill or repulsed.

Utilitarianism focuses on the consequences or effects of actions. The German philosopher Immanuel Kant took a different approach to ethics. Kant focused not on the consequences of actions but on the intentions with which they were undertaken. Moral actions, Kant stipulates, are those undertaken out of a sense of duty. His "categorical imperative" states an unconditional ethical duty always to "act only according to that maxim by which you can at the same time will that it should become a universal law." That means that we should always act in a way that others might replicate without the aggregation of all these actions undermining the conditions that make them possible.

For instance, if the maxim "Tell a lie whenever doing so well serves your purposes" were universally adopted, nobody would trust anyone. In such a world, lying would no longer serve our (self-interested) purposes because no one would believe what we say in the first place. The maxim "Always tell the truth," in contrast, can become a universal law. This maxim may not be easy to follow in all circumstances. But were everyone to abide by it, telling the truth would still produce effective communication.

The categorical imperative, like the utilitarian pleasure principle, is wonderfully clear and concise. As one might expect, however, it becomes challenging to implement. In a world where diversity is crucial to resilience, universalizable forms of behavior are not necessarily optimal. We would not want everyone to make our specific career choices, for instance: imagine a world in which everyone became a plumber, or a lawyer. So the maxims that guide action must remain very general, lest they bring everyone to do the same thing, at the same time, in the same way. But this requirement of generality means that the categorical imperative provides little guidance for our daily activities.

Still, Kant's categorical imperative lends support to a sustainability ethic. It suggests that we have an obligation to live in a manner that everyone might replicate without depleting, degrading, or otherwise undermining the social, environmental, and economic conditions that make such lives possible.

The American philosopher John Rawls explicitly extended a Kantian ethics to include responsibilities to future generations. He captured the spirit of the Golden Rule and the categorical imperative with his principle that we should always act as if a "veil of ignorance" prevented us from knowing what race, religion, or nation we belonged to, whether we were healthy or ill, fit or disabled, rich or poor, members of the current or a future generation.[11] Unaware of our abilities or propensities and our socio-economic, demographic, geographic, and generational status behind the veil of ignorance, our choices and actions would be fair to all stakeholders, for, as far as we know, we might be any one of them. By this account, a Rawlsian would have a small ecological footprint, embrace social empowerment, and reject corrosive disparities in wealth.

Whether our pursuit of sustainability is structured by a utilitarian principle that seeks to maximize good consequences for the greatest number, a Kantian or Rawlsian principle focused on fulfilling our obligations, or any other principle, it will rely on moral values. For many people, moral values are grounded in religious beliefs. Just as various versions of the Golden Rule were articulated millennia ago within diverse spiritual traditions, so ethics today is frequently developed within religious communities. Fostering an ethic of sustainability may require the aid, and perhaps transformation, of the religious traditions that provide moral instruction to so many.

The historian Lynn White, Jr, first made this claim in an article published in the journal *Science* in 1967. In "The Historical Roots of Our Ecologic Crisis," White argued that Christianity bore a significant "burden of guilt" for many of the environmental problems then facing society.[12] Christian scripture, such as Genesis 1:28, portrayed human beings as the "masters" of nature with the God-given prerogative to subdue it. This religious narrative establishes humanity's right, and even duty to dominate the natural world. It is precisely this "old story, the account of how the world came to be and how we fit into it," that required radical revision, according to Father Thomas Berry.

Subsequent to the controversy that White's article pro-
voked, many westerners came to believe that indigenous spiritual traditions or eastern religions such as Buddhism might
provide "eco-friendly" alternatives. In turn, efforts were
made to explore and emphasize the environmentally attuned
features of Christianity, and those of other world religions
such as Judaism, Islam, and Hinduism.[13]

Religions are often grounded in sacred texts. But these texts
– typically written over hundreds of years by many authors –
do not present a single story. Rather, they interweave diverse
narrative strands. As such, they require ongoing interpreta-
tion and may serve as inspiration for a wide array of convictions and practices. White himself, while celebrating the
non-dualism of pagan and Asian spirituality, also found within
the Christian tradition resources for an ethic of environmental
concern. He dubbed St Francis of Assisi the "patron saint of
ecology." In recent years, religious leaders and organizations
from various traditions have interpreted their scriptures to
align with environmental concerns and have incorporated
sustainability into their sermons and creeds. The greening of
religion is widespread.

Religions, like other forms of cultural life, are living traditions. Along with biological life, they evolve over time, adapting to circumstances by building on their inheritance. What
Leopold said of ethics, therefore, applies equally to the values
fostered by religious traditions and to religious traditions
themselves. Ethics, Leopold maintained, are not unchanging
rules written in stone. Rather, they are "tentative" products
of "social evolution." Our moral traditions remain fluid
"because evolution never stops."[14] Ethics are evolving sets of
principles and rules that guide human behavior within an
ever-changing world. Religious traditions, which both reflect
and serve as sources for ethical beliefs and practices, also
evolve.

The earliest spiritual and religious practices were versions
of pantheism, animism, and polytheism. Here, nature as a
whole, or a set of natural entities, was taken to be divine.
Particular deities were believed to inhabit natural objects or
forces. The greening of contemporary religions might be seen
as a rediscovery of this ancient relationship between nature

and the divine. Virtually all religions view Creation as sacred, as arising from the hand(s) of god(s). And most religions have developed some notion of a "stewardship ethic" that depicts human beings as trustees obligated to the caretaking of the natural world. In turn, all religions advocate various forms of social justice and equity, often exemplified by their particular version of the Golden Rule. It follows that religion, as a primary source of moral values for many people, can and likely will play an important role in the development and dissemination of sustainability ethics. Pope Francis's 2015 encyclical addressing climate change and planetary protection, "On Care For Our Common Home," provides a recent example.

Aesthetics and education

For Leopold, whether an action is right or wrong depends on its effects on the community of life. He argued that one should "Examine each question in terms of what is ethically and esthetically right, as well as what is economically expedient. A thing is right when it tends to preserve the integrity, stability, and beauty of the biotic community. It is wrong when it tends otherwise."[15] Notice that aesthetics stands side by side with ethics as a criterion for right action. The beauty of the biotic community, not only its capacity to serve our economic interests or even its own long-term viability, is a central concern.

Rachel Carson insisted that the attraction to and preservation of "natural beauty has a necessary place in the spiritual development of any individual or any society." Carson continued: "I believe that whenever we destroy beauty, or whenever we substitute something man-made and artificial for a natural feature of the earth, we have retarded some part of man's spiritual growth."[16] Like Leopold, Carson celebrated the beauty of nature and the desire to safeguard it. Without this aesthetic sensibility, perhaps grounded in biophilia, the protection of the environment and the development of an environmental ethic would be difficult, if not impossible.

The aesthetic appreciation of nature is more than a means to stimulate environmental protection. It is a lifeline to a rich and meaningful life. Carson wrote that:

> Those who dwell, as scientists or laymen, among the beauties and mysteries of the earth are never alone or weary of life. Whatever the vexations or concerns of their personal lives, their thoughts can find paths that lead to inner contentment and to renewed excitement in living. Those who contemplate the beauty of the earth find reserves of strength that will endure as long as life lasts.[17]

The beauty of nature helps us maintain psychological, spiritual, and physical health.

In her book, *Healing Spaces*, Dr Esther Sternberg has explored the power of natural beauty to heal the body. Studies demonstrate that hospital patients recuperating in rooms that offer a view of nature heal faster than those patients who have no visual connection to the outside world. Sternberg argues that our exposure to nature's serenity and beauty has a direct impact on our levels of anxiety, stress, and the resilience of our immune system.[18] We have long understood that a degraded environment is not good for human health. We now know that this relationship occurs not only by way of the toxins that we breathe, drink, eat, or absorb through our skin. The health of our bodies and minds is also strongly affected by what we absorb through our senses of sight, smell, and hearing.

Greening urban landscapes reduces ambient urban air temperatures in the summer, sequesters carbon, and produces cleaner air. It also stimulates people to become actively involved in nature conservation. But there is another reason to build "biophilic cities" that include trees, parks, greenways, riverfront walks, easy access to natural surroundings, and gardens growing on rooftops, balconies, yards and neighborhood lots. Direct experience of the natural world markedly improves physical and mental health. It heightens residents' moods, makes for safer, better neighborhoods, and stronger community relationships.[19]

With this in mind, consider the fact that youth today engage in much less outdoor activity than ever before. Children in the

United States, for instance, spend about half as much unstructured time outdoors as their parents did, and a quarter as much as their grandparents.[20] And they spend more time in front of television, computer, or video-game screens than in any activity other than sleep: as much as 40 hours a week.[21] The effects on brain development and health are noticeable.[22] In turn, much of the time spent in front of electronic media exposes youth to intensive advertising. The average child sees more than 40,000 advertisements each year. Consequently, young people today recognize over a hundred corporate logos for each plant or animal they can identify that is native to their locales.[23]

In his path-breaking book, *Last Child in the Woods*, Richard Louv maintains that children who regularly interact with nature receive a powerful form of therapy that has patent health and psychological benefits. Coining the phrase "nature-deficit disorder," Louv argues that the current pandemic of childhood attention and hyperactivity disorders, not to mention childhood obesity and diabetes, is best combated by reintroducing youth to the pleasures of exploring earthscapes and ecosystems.[24] Seeing, hearing, and touching nature every day, he maintains, is healthy and healing. In turn, such activities foster a love of the natural world, and this devotion to things natural is crucial if future citizens and consumers are to commit themselves to environmental protection.

Having evolved within natural environments that both sustained and pleased, the love of nature became a deep and enduring part of human being. While our contemporary urbanized landscapes, apartments, offices, and video monitors often keep us separated from the natural world and its beauty, our species is still subject to the genetic inheritance that predisposes us to thrive in direct contact with nature. To a significant extent, this predisposition is an aesthetic sensibility, an appreciation of natural splendor. Our physical senses evolved through direct participation in diverse webs of life, and our biophilia, to large extent, is an appreciation of the beauty of nature.[25] Nature has always been the primary source of inspiration for artists. They limn the beauty and bounty of the earth on canvas, in wood or stone, in music, dance, design, and poetry. Even in our homes and workplaces, which serve as refuges from the outside world, we incorporate natural objects and pieces of art inspired by nature. And those who can typically install large windows in their homes so they might gaze upon oceans, lakes, rivers or streams, trees, shrubs, grasses, or flowers. Seeing nature stirs our souls.

Culture is often opposed to nature. It is understood as an antagonist and alternative to natural impulses, something that allows us to rise above our barbaric past. This opposition has spawned the famous "nature versus nurture" debate. Nature is what is given, the product of innate capacities, instincts, genes, and DNA. In contrast, culture is what is developed, the product of human values, morals, customs, and institutions. But this is a false dichotomy. Culture is the way human beings express their nature. It is natural for us, given our genetic makeup, to develop culture. And cultural life, at its best, is the cultivation and celebration of natural gifts.

As the realm of art and aesthetics demonstrates, nature and nurture can be the staunchest allies. The former makes possible, informs, and inspires the latter. As painters, sculptors, poets, singers, musicians, dancers, and designers, we pursue the beautiful as much if not more than the instrumentally useful. Aesthetic pursuits lend both pleasure and meaning to our lives. Without them, at least for many, life would be intolerable. And without the natural world to inform and inspire our aesthetic sensibilities and pursuits, much if not all art would be impossible.

The practice of sustainability is not the pursuit of bare existence. It is the pursuit of the good life for an ever-expanding community. Absent aesthetic endeavors, emotional attachments, and spiritual experiences, human existence might still be possible. But the good life surely would not. Nature sustains us ecologically and economically. It nourishes us as biological creatures and provides resources for our daily lives. Nature also sustains us aesthetically, as the inspiration and model for our artistic endeavors. And it sustains us emotionally and spiritually, as a source of inspiration and meaning.

The practice of sustainability entails the preservation of the natural world in all its beauty, and the cultivation of the aesthetic, emotional, and spiritual sensibilities that allow us to gain pleasure and meaning from it. All this is made possible through education. A culture of sustainability, like any other culture, establishes and maintains itself through teaching and learning.

Societies educate their young, and re-educate their adults. This allows a culture to carry on traditions and practices and adapt to a changing world. The word "education" derives from the Latin word *educare*, which means to bring up and lead forth. To educate is to bring up and lead forth the young – and the young at heart – into the world. Education is an introduction to the ecological systems composing the natural world, and to the beliefs, values, practices, relationships, and institutions of the social world. It entails gaining knowledge and skills so the natural and social world might be well navigated, carefully preserved, and beneficially transformed.

But education is a double-edged sword. Learning is not always put to the best use. All too often, acquired knowledge and skills are employed to exploit, degrade, and destroy the natural world, other human beings, and cultural achievements. It is said that the pen is mightier than the sword, that learnedness is more powerful than brute strength. But history often demonstrates that pen and sword, brain and brawn, work in tandem to gain and use power and wealth in socially and ecologically unsustainable ways.

For our hominin ancestors, preserving their culture and natural environment may have been as simple as passing along to progeny a small handful of skills and values. Culture was very limited; instinct did most of the work. And the surrounding ecosystems, inhabited by a small population with rudimentary technology, were rarely if ever threatened. Today, the situation is very different. Neither our cultural world nor our natural environments can be preserved, or beneficially transformed, without extensive education. And with each passing day – as human numbers, technological development, and ecological degradation grow – educating for sustainability becomes ever more pressing.

The study of nature is perhaps the oldest form of education known to humankind. Since ancient Greek and Roman times, instructional texts focused on gardening, horticulture, agriculture, and animal husbandry. Education aimed at conserving specific natural resources, such as forests and wilderness, date back at least to the nineteenth century. Only in the last few decades, however, were general educational programs developed that were designed to foster the knowledge, values, and skills required for the conservation of resources and the preservation of the natural environment. The first Earth Day on April 22, 1970 – conceived by US Senator Gaylord Nelson of Wisconsin as a national "teach-in" about environmental issues – was the impetus for the development of popular environmental education programs. Later that year, the National Environmental Education Act established the US Office of Environmental Education, with a mandate of awarding grants to develop curricula in environmental education and to provide teachers with professional development. Senator Nelson's home state of Wisconsin became the first to mandate environmental education for years K through 12. A majority of other states followed suit.

Internationally, environmental education experienced a similarly rapid development. The United Nations 1972 Conference on the Human Environment held in Stockholm deemed environmental education crucial to grappling with the world's most pressing problems. Following the Stockholm conference, other international meetings, such as the Tbilisi (Georgia) Intergovernmental Conference on Environmental Education of 1977, outlined specific objectives and guiding principles for environmental education, while re-emphasizing its importance.

By the late 1980s, the international promotion of environmental education began to dovetail with the growing support for sustainability and sustainable development. The Talloires Declaration, composed at a 1990 conference in Talloires, France, was the first official statement by university administrators outlining a commitment to sustainability in higher education. It provided a 10-point plan of action for making environmental literacy and sustainability part of the mission of institutes of higher education. The University Leaders for a Sustainable Future (ULSF) serves as the official Secretariat of the Talloires Declaration. The organization's mandate is "to support sustainability as a critical focus of teaching, research, operations and outreach at colleges and universities worldwide through publications, research, and assessment."[26] The Talloires Declaration has been signed by more than 400 university presidents and chancellors hailing from over 50 nations.

After the Rio Summit of 1992, education for sustainability became widely endorsed at the governmental level. Agenda 21, the sustainability program for the twenty-first century signed at Rio by 178 nations, specifically endorsed "reorienting education towards sustainable development." Chapter 36 of Agenda 21 proposes specific goals, methods, and resources for sustainability education. With a mandate of implementing these proposals, an association of European universities established the University Charter for Sustainable Development and launched Project COPERNICUS (Co-operation Programme in Europe for Research on Nature and Industry through Coordinated University Study).[27] Its projects are active at more than 300 universities in 37 European countries.

Education for Sustainable Development, also known as Education for Sustainability or Sustainability Education, is now extensively promoted internationally and has been implemented within thousands of primary, secondary, and post-secondary institutions. It takes myriad forms: sustainability issues are integrated into existing courses across the curricula; new courses are created to introduce students to sustainability or tackle specific sustainability concerns; and wholly new programs, including concentrations, minors, majors, and graduate degrees in sustainability have been developed. These programs tend to be strongly interdisciplinary, exposing students to diverse forms of inquiry and challenging them to integrate fields of study that otherwise might remain isolated.

A decade after writing *The Limits to Growth*, Donella Meadows observed that "The world is a complex, interconnected, finite, ecological-social-psychological-economic system. We treat it as if it were not, as if it were divisible, separable, simple, and infinite. Our persistent, intractable, global problems arise directly from this mismatch."[28] Sustainability educators take on the task of teaching beyond the narrow disciplinary boundaries that often structure academic life in order to equip students for the complexities of an interdependent world.

To mark the inroads that sustainability studies have made into the world's educational institutions, and to underline the vast amount of work yet to be done, the United Nations General Assembly declared 2005–14 the Decade of Education for Sustainable Development (DESD). It tasked its Educational, Scientific and Cultural Organization (UNESCO) to implement related programs and goals. Within the DESD framework, sustainability education is not restricted to what happens in formal institutions of learning, or to programs launched in response to governmental initiatives. It is a much more widespread and diverse phenomenon carried on within civil society and workplaces.

The United Nations Environment Programme (UNEP), which was created in 1972 following the Stockholm conference, has a mandate of promoting this education and training. Its mission is to develop "understanding, skills and values that will enable people to participate as active and informed citizens in the development of an ecologically sustainable and socially just society."[29] These skills, values, and understanding include: critical thinking and interdisciplinary knowledge; understanding systems and complexity; appreciating multistakeholder perspectives and developing empathetic capacities; planning and managing change; clarifying values, communicating effectively, and making decisions under uncertainty; democratic participation, consensus building, and cooperative action.[30]

Sustainability, whether in formal educational institutions or in civil society, is less a discrete subject matter than a particular approach to learning.[31] To be sure, sustainability educators attempt to convey data about important subjects. But more importantly they promote learning about the interdependencies that characterize the social, economic, and ecological realms at local, national, and global scales. They underline the complex interactions that occur within and between these systems. And they encourage transparent systems of knowledge and power within which interdisciplinary learning and effective action can occur. The cornerstone of a culture of sustainability is expansive, interactive, and adaptive learning.

Core values and cultural creativity

It would be a much simpler world if increased information and knowledge always shifted opinions and attitudes, if changed opinions and attitudes always transformed values, and if transformed values always altered behavior. But the world is not simple. And no part of the world is more complex than the human mind.

Studies consistently demonstrate that the acquisition of knowledge does not often or easily translate into changes in behavior. Even when opinions, attitudes, and values are transformed on account of knowledge gained, changes in behavior tend to lag far behind, if they follow at all. Indeed, increased learning may become a substitute for action rather than a catalyst for change.[32] In such cases, the gaining of information becomes an end in itself rather than a means to reorient behavior. And since there is always more information to be gained, action can be endlessly deferred. Changing one's mind is easier than changing one's life, and often becomes a substitute for it.

Knowledge of sustainability-related issues in the general public is neither widespread nor particularly deep. There remains much ignorance regarding the complex ways in which ecological health, economic welfare, social empowerment, and cultural creativity are threatened, can be pursued, and interrelate. Even when levels of awareness and concern for such things are significantly increased, behavioral change lags far behind.[33] In short, developing sustainability knowledge and values and translating them into enduring practices is difficult.

Effective sustainability education – education that produces behavioral change – both informs and empowers. Without a sense of responsibility for the welfare of present and future generations, people will not be motivated to care for anything beyond their own backyards. Without investigative and problem-solving skills, people will not understand or know how to address threats to their or others' welfare. And without the capacity to effect meaningful change, both as individual agents and as members of communities, people will not take action.[34]

Fearmongering or scare tactics are seldom helpful. The world can be a frightening place, and to be educated is to understand the gravity of current conditions and the formidable challenges that lie ahead. Viewing the world through rose-tinted glasses is not helpful. But a "doom and gloom" approach to sustainability education is a recipe for failure. That is because fear is a double-edged sword. It can induce appropriate action in certain circumstances, namely when the threat is clear and an effective response is straightforward. However, when threats and effective responses are complex, fear can be psychologically debilitating. It can lead people to ignore the problem at hand, burying their heads in the sand; or to simply deny that the problem exists in the first place; or to become cynical or apathetic because they see no solutions. If you find yourself in a burning building, then fear will quickly get you headed for the door. Here, a clear threat produces a simple, effective response. But if you find yourself on a quickly warming planet, fear by itself will not generate effective action.

To be effective, sustainability education must not only inform by way of facts. It must foster a sense of responsibility, develop investigative and problem-solving skills, and cultivate the experience of empowerment or agency. Only in this way will complex threats be met with informed, collaborative, effective responses. Sustainability education faces a unique challenge in this regard. Noting that the best educated in society – namely those who graduate with degrees from the most prestigious institutions of higher education – are generally the most profligate consumers of precious resources, David Orr writes that "It is not education, but education of a certain kind, that will save us."[35] The kind of education that Orr has in mind develops investigative and problem-solving skills and instills a sense of responsibility and empowerment.

Environmental educator C. A. Bowers writes of the need for a curriculum that "provides both the experience of being an interdependent member of a human/biotic community and an understanding that interdependence is the basic relationship that connects past, present, and future generations."[36] What is required is a "schooling for interdependence" that encourages and empowers people to understand, navigate, safeguard, and transform the connections that define their world.[37]

An education in interdependence is inherently difficult to provide in a formal institutional setting. For the most part, our schools and institutes of higher education are places students go to learn, not to live. Yet the breadth and depth of our lived interdependencies are most fully experienced in the nested communities that daily sustain us.

Bioregionalists have long argued that we only protect what we love, and that we only love what we know. Learning about the community of life, they hold, is best achieved from the inside, as an active participant. This learning not only provides information and knowledge, but also cultivates care, responsibility, and a sense of agency. The most successful education in sustainability, it follows, will not only inform students but also help them become active participants in the nested communities that sustain them. This may take the form of community internships, experiential interactions with local flora, fauna, and culture, and other means of instilling a "sense of place."

Direct interaction with nature is educational and therapeutic and instills a commitment to conservation. Likewise, direct interaction with people is crucial to the cultivation of empathy and a commitment to social equity and empowerment. Studies documenting the decline of fellow feeling among youth are disturbing in this regard.[38] Quite possibly, the time young people spend in front of televisions, computer screens, playing electronic games, and surfing the internet has produced not only a loss of relationship to the natural world but increasing social isolation as well. Getting students to interact regularly with nature while becoming actively involved in their social communities is a core challenge of a sustainability education.

In many respects, sustainability is a "lifestyle" movement. With this in mind, its advocates ask us to "reduce, reuse, and recycle," and to embrace "voluntary simplicity." Organizations such as Adbusters promote an annual, international "Buy nothing day" in an attempt to transform consumer culture – a culture where shopping provides the most prominent pastime for affluent people and is even touted as a form of psychological self-help known as "retail therapy." The project has been criticized for ignoring the plight of the billions of poor people for whom buying nothing (to eat, drink, or wear) is a regular and debilitating burden. Still, such efforts underline the sad fact that shopping, rather than intellectual, artistic, ethical or spiritual engagements, may be the defining feature of our culture. When consumption is no longer simply a means of satisfying basic human needs but has become an end in itself, the pursuit of the true, the beautiful, and the good takes a back seat.

Around 20 years ago, books such as *50 Simple Things You Can Do to Save the Earth* were popular. Its pages addressed changes that individuals might make to their lifestyles. The challenges involved in shaping public policy and law and restructuring economies were mostly ignored. Changing lifestyles is crucial but insufficient. It remains only part of a broader project of cultural change that includes the transformation of educational, ethical, religious, artistic, social, political, legal, and economic practices and institutions at local, national, and global scales.

While sustainability will always require attentiveness to the impact of technology, many of the most pressing problems we face today do not have technological solutions. They require cultural change. Consumption for consumption's sake is not a sustainable practice, and its remedy is largely cultural. Even problems that seem to beg for a technological solution, like high fertility rates, cannot be adequately addressed without cultural change. Making contraceptive devices available is a necessary, but far from sufficient, means for addressing overpopulation. The absence of effective education programs along with certain social and religious norms are the biggest barriers to effective family planning in many countries. Keeping girls in school and giving women greater control of their lives are the most effective means of reducing fertility rates here.[39]

The challenge of creating a "culture of sustainability" is monumental. Cultures are enacted in stories populated with particular themes, values, and protagonists. The "frontier" culture that developed in the United States as the land was being settled by Europeans, for instance, celebrated rugged pioneers who tamed a continent, bringing it under axe, plow, drill, and gun, transforming it from wilderness into a resource for national industrial power. But, as Thomas Berry observed, we need a new story today. A culture of sustainability has its own themes and values, and its own heroes.

The activist Wangari Maathai (1940–2011) is a case in point. Maathai won the 2004 Nobel Prize for her efforts with the Kenyan Greenbelt Movement. She struggled against a culture of patriarchy most of her life, overcoming many personal barriers in her efforts to organize and employ women in her native Kenya to plant trees. Since Maathai began the movement in 1977, tens of thousands of women have been trained and over 30 million trees planted across Africa. The United Nations Environmental Programme created its own Billion Tree Campaign in response to these heroic efforts. But Maathai's tree-planting movement was more than an act of reforestation, erosion control, and conservation. It underlined the capacity of average citizens to take control of their lives and land, it fostered equity, and it enacted a sense of responsibility to future generations. It was sustainability leadership at its best.

In his popular novel, *Ishmael*, Daniel Quinn tells a tale of a gorilla from a zoo who has learned to speak. The gorilla, named Ishmael, seeks a student who wants to save the world in the face of humanity's unsustainable ways. Ishmael states that a culture is a people who are enacting a story. The primary human story, our "living mythology," is the tale of the Takers. This story begins with the agricultural revolution, as humans gained technological control of the earth and increasingly exploited it to meet their needs and growing wants. As control and exploitation grew, so did human numbers, the depletion of natural resources, and the extinction of species. Takers assume that the world belongs to them; its domination is their birthright. Because Takers believe that all natural resources exist for their benefit, their culture promotes the endless growth of human desire. The result is despoliation of the earth.

But there is an alternative story. Before the Taker narrative gained hegemony, a culture of Leavers had developed. Leavers do not believe that the world belongs to them. They believe that they belong to the world. Leavers did not produce surplus food through agriculture, or surplus power through hierarchical social structures. They lived within ecological limits, as equal participants of a diverse web of life.

Human beings, as cultural creatures, always live out one story or another. So unless another narrative is found to replace the Taker ideology, Ishmael observes, there is no hope for the earth. While there is much to be learned from Leaver culture, Quinn suggests, we cannot revert to their primitive ways of hunting and gathering. We cannot undo all of the knowledge and skills and evolutionary development that have made human beings self-conscious creatures who plan and produce their own welfare with the help of technology. Rather, a new cultural story has to be developed, one grounded in our species' participation in an ongoing evolutionary odyssey with a vast diversity of other creatures. This story must integrate the incredible technological power we have come to develop with a newfound or rediscovered sense of ecological community and responsibility.

Sustainability provides such a narrative. It does not propose that humankind can or should retreat to a primitive way of life. That would be idle romanticism. In any case, the point is not to abandon our cultural, social, political, economic, and technological knowledge and skills. We need not deny our unique capacities and aspirations. But neither should we relate to the earth chiefly as its plunderer. The story of sustainability is a living mythology that charts our creative responsiveness to an interdependent world.

A key theme of the story of sustainability is the safeguarding of social, cultural, and natural diversity. Diversity grounds resilience. It allows for better adaptation. If the only tool we have in our toolbox is a hammer, we'd better hope that all we ever encounter in life are nails. In contrast, if we have a wide variety of tools, any number of problems can be adaptively addressed. And if we understand that not all problems have technological solutions, we are well on our way to developing a culture of sustainability.

Culture is typically seen as something to be preserved, to be passed on intact from generation to generation. Often it is. But to survive in the long term, a culture must adapt. Sustainability entails conservation. But to be an effective force for the conservation of core values and relationships, sustainability must also be a force for change.

Consider the core values of nutrition and health. Our natural cravings for things sweet and fat were adaptive traits over the first few million years of our development as a species. These cravings helped us secure the core values of nutrition and health. During this time, things sweet were relatively hard to find and often out of reach. Things fat had a habit of hiding and running away. It was difficult for early hominins to overindulge their natural inclination to ingest sugars and lipids. The amount of energy expended in gathering and hunting stood in relative balance to the amount of energy derived from the sweet and fat things that could be picked and captured. In the Pleistocene, obesity was not a problem.

The agricultural revolution, followed by industrial agriculture and food production, changed all that. Now things sweet and fat became available on every street corner, and they fill our cupboards. Without scarcity to limit intake, we have to rely on our self-control to prevent overconsumption. What was once an adaptive trait – the natural inclination to seek out sweets and fats – has today become a very maladaptive trait, producing widespread obesity, rampant diabetes, and heart disease. Our core values of nutrition and health are no longer well served by the unchecked propensity to eat everything sweet and fat within reach.

We might be able to genetically redesign our bodies so that we no longer desire or absorb sweets and fats. But this technological solution would undoubtedly have noxious unintended consequences. Better to solve for pattern, adapting our beliefs and behavior – and the institutions that structure and channel these beliefs and behaviors – to secure multiple benefits. The architecture of choice might play a pivotal role here, helping us adapt to changing circumstances. Only an adaptive culture can conserve its core values and relationships.

In cultural life, as in biological life, continuity and change are partners in a dance. Sara Parkin succinctly captures this dynamic when she writes that "Sustainability leadership is about the continuing participation of our species in evolution. Get it wrong and we are fossils."[40] Sustainability provides a narrative that describes our "co-evolutionary" participation in the web of life.[41] We are not only changing the world. We are an interactive part of a changing world, partners in a co-evolutionary dance. The point is to ensure that the music never stops.

Inquire and Explore

1. What are the key features of a sustainability education?
2. Does sustainability have a spiritual dimension?
3. How can the pursuit of sustainability stimulate your creativity?

******

- Investigate the ethics of sustainability.
- Become inspired and take action to create a more sustainable world.

 For web resources and more information on these topics, please visit conservationandcreativity.net

Conclusion

To be sustainable, a practice, relationship, or institution must maintain the health of the natural environment. It must be economically sound. Its participants must be empowered and equitably treated. And it must foster a creative culture of adaptive learning so that the scale and speed of change can be well managed. Only the balanced pursuit of these four goods in the context of increasing global interdependencies ensures that present needs can be met without sacrificing future welfare.

Our exploration of sustainability as a concept and as a practice underlines the need to expand our time horizons, extending our responsibilities to future generations while developing an appreciation of our social and ecological inheritance. Expanding our spatial horizons is equally important. It helps us appreciate the global commons, address the social and ecological costs of poverty and inequity, and carefully assess the benefits and dangers of globalization – the steady expansion of social, technological, political, economic, and cultural interdependencies.

Knowing that the resilience of ecosystems is grounded in their biodiversity, and that a healthy environment has both intrinsic worth and provides crucial resources and services, we may better appreciate the importance of nature conservation. Green technology can help us protect nature while serving crucial social needs. At the same time, technology produces unintended consequences. And not every problem can be well addressed with a technological solution.

Indeed, many of the most important challenges we face are not technological but political and legal. Hard work in these arenas is required to ensure that policies, regulations, and laws help, or at least do not hinder, the building of sustainable societies. The more equitable sharing of knowledge and power is a crucial part of this task. While national governments are mandated with protecting their citizens, achieving security today increasingly depends on global caretaking and the fostering of self-reliant communities that embrace the responsibilities of governance.

The sustainability of communities no less than that of nations rests on strong economic foundations. Sustainable businesses cannot be based on unrestricted and unending growth. Ecological economics fosters needed social and technological development without raising levels of consumption and waste beyond the carrying capacity of the planet. It is difficult to exaggerate the importance of greening business. But, at the end of the day, sustainability is as much a cultural achievement as an economic affair. Educating our intellectual, aesthetic, ethical, and spiritual sensibilities will play a large role in crafting a sustainability narrative that integrates economic welfare with ecological health and social empowerment. Sustainability is an adaptive art wedded to science in service to ethical vision.

We now stand at a crossroads. In the past 50 years, child mortality rates across the globe have declined by two thirds, while life expectancy has increased by almost 20 years. Literacy rates and per capita income have also increased, while the percentage of people living in extreme poverty has decreased. That is certainly progress for our species, and we must give credit to technology and economic growth for these benefits. During this same half-century, however, the impact of human enterprise and human numbers on the earth has been nothing less than devastating. Population has more than doubled, increasing by 4 billion people. Carbon dioxide emissions have tripled, paper consumption has quadrupled, fertilizer use has quintupled, and the consumption of other natural resources has also increased at patently unsustainable levels. Citing evidence for peak oil and observing the effect of burning fossil fuels on the climate, Richard Heinberg observes that:

> There are a few hopeful indications that a shift toward sustainability is beginning. But there are also discouraging signs that large political and economic institutions will resist change in that direction. Thus, the most likely trajectory for the energy transition will consist of the collapse of industrial civilization as we know it, probably occurring in stages over a period of several decades.[1]

For Heinberg, the writing is on the wall. If we do not bring our economies and lifestyles back into balance with the planet's carrying capacity, ecological devastation and social disorder will do the job for us – brutally and with tremendous costs in human lives, cultural achievements, social relationships, and ethical values. History indicates that collapse was a fate suffered by scores of once successful societies. And while a look to history should help us appreciate the frequency of collapse, we also know that the scale of today's crisis is unprecedented. "Every time history repeats itself," Joseph Tainter observes in his study of the rise and fall of complex societies, "the price goes up."[2] Today, the stakes in the game are planetary in scope.

Stocks of natural resources are precipitously falling and pollution sinks are steadily filling. Some of these sinks – like the atmospheric and oceanic sinks we use for carbon dioxide – are already overflowing. We are living at a time of peak consumption and peak waste. Collapse *is* a real possibility. But that is no reason for despair. Indeed, despair is the last thing we need, for it leads to resignation and passivity.

Our connected world admits no bystanders. To live sustainably is to take responsibility for our lives and our communities. Dire economic, social, ecological, and cultural problems are best approached as opportunities to exercise creativity and compassion. Averting collapse will necessitate sacrifice. But there is mounting evidence that personal happiness, social equity, and strong community can grow together as consumptive, materialistic lifestyles wane.[3] We are living at a time of peak consumption and peak waste. But we are not even close to hitting peak prosperity. There is good reason for hope.

Sustainability demands the balancing of multiple goods whose pursuits are generally compatible, and often mutually reinforcing. But the practice of sustainability has to be grounded in pragmatic realism: all goods cannot be simultaneously maximized; anything in excess, including the pursuit of ideals, eventually proves counterproductive; nothing is without its inherent dangers; and all of our actions (and non-actions) have unintended consequences. The time for silver bullets and panaceas has passed. Once-and-for-all solutions do not exist for the problems we face today. Still, there are better and worse ways to respond.

Better responses do not address problems in isolation. They meet complex, connected problems with synergistic responses. They solve for pattern.

Better responses do not address responsibilities in isolation either. To play our part fully in the unfolding story of sustainability, we must respond not only as responsible consumers but also as responsible stakeholders and citizens. Changing light bulbs or even lifestyles is not enough, unless lifestyle changes include becoming more socially and politically active. Better responses occur sooner rather than later. The greatest part of practical wisdom consists in being wise in time.

Finally, better responses vary with the circumstances. One size does not fit all. The only sure thing we can count on in this life is that we will be surprised by what comes round the corner. In such a world, the only good response is an adaptive one.

Societies that balance sound economies with social empowerment, ecological health, and cultural creativity are primed for adaptation. But that does not mean everything can or should change at once. Changes to the biosphere, and our own transformation as individuals, communities, societies, and as a species, have to occur at an appropriate scale and speed. Only in this way is resilience assured through the conservation of core values and relationships. Sustainability is Janus-faced and two-handed. It is future-focused but with an eye to its inheritance. And while it deftly embraces adaptive change with one hand, it also firmly grasps the need to conserve. Sustainability blends creativity with conservation:

the conservation of nature, the conservation of resources, and the conservation of options.

Core values and relationships are not immune to change. While we might think of international human rights as a core value today, it would have been nonsense to Pleistocene hunter-gatherers or ancient Sumerians. Still, core human values – such as truth, beauty, and moral goodness, and more specifically, compassion, economic opportunity, justice, liberty, physical security, cultural creativity, and a vibrant natural environment – are enduring. They contribute to a high quality of life and define human prospects. If these values are widely pursued and maintained, individuals and societies are likely to flourish.

The goal is to build communities and societies that are prosperous enough to preserve their core values and relationships. In the recent past, prosperity and sustainability were often at loggerheads. The paleo-prosperity of the last few centuries entailed maximizing the consumption of energy and natural resources. Neo-prosperity, the type required today, will be generated through the life-enhancing, community-building, health-producing, culturally creative means of satisfying our needs and reasonable wants.

Neo-prosperity entails technological ingenuity. It also entails shifting our values and reward systems. A top hedge-fund manager in 2008, who probably shared a portion of the blame for that year's financial meltdown and the ensuing global recession, earned a salary equivalent to that of 20,000 schoolteachers – the people we trust to instruct and mentor children so they might carry on civilization and safeguard the biosphere. There is much work to be done – personal and political – if we are to create a culture of sustainability. Much of that work falls within the realm of education, as we learn about the interdependencies that characterize our world. The creation of sustainable societies depends upon adaptive learning grounded in transparency and empowerment. Translating education into action and infusing cooperative action with learning are crucial tasks.

Sustainable societies will not be utopias. Not all of our wants can be satisfied. Every adult should know that. Making sacrifices for worthy things is a sign of maturity. It is intrinsic to the good life. With this in mind, the pursuit of sustainability can both inspire and discipline us. Like musicians practicing their instruments, or athletes practicing their sports, people practicing sustainability are in for some hard work. But there is much joy to be experienced as well. Indeed, like fine musicians or athletes who relish their activities, only those delighting in the practice of sustainability are likely ever to become very good at it. Practicing sustainability can be one of the most rewarding things we do – an opportunity for discovery, personal fulfillment, enriched community, and celebration.

Notes

Introduction

1 John D. Mittelstaedt, Clifford J. Shultz II, William E. Kilbourne, and Mark Peterson, "Sustainability as Megatrend: Two Schools of Macromarketing Thought," *Journal of Macromarketing* 34(3) (2014): 253–64.

2 David Owen, *The Conundrum: How Scientific Innovation, Increased Efficiency, and Good Intentions Can Make Our Energy and Climate Problems Worse* (New York: Riverhead Books, 2011), p. 246.

3 David Selby, "The Firm and Shaky Ground of Education for Sustainable Development," *Journal of Geography in Higher Education* 30(2) (July 2006): 363.

4 Liz Ford, "Rio+20 Politicians Deliver 'New Definition of Hypocrisy' Claim NGOs," *Guardian*, June 21, 2012, accessed at http://www.guardian.co.uk/global-development/2012/jun/21/rio20-politicians-hypocrisy-ngos?intcmp=239.

5 Donald Worster, "The Shaky Ground of Sustainability," in George Sessions, ed., *Deep Ecology for the 21st Century* (Boston, MA: Shambhala, 1995), p. 418.

6 Peter Blaze Corcoran, Kim E. Walker, and Arjen Wals, "Case Studies, Make-your-case Studies, and Case Stories: A Critique of Case-study Methodology in Sustainability in Higher Education," *Environmental Education Research* 10 (2004): 7–21; esp. 8–9. See also Arjen Wals and Bob Jickling, "Sustainability in Higher Education: From Doublethink and Newspeak to Critical Thinking and Meaningful Learning," *Higher Education Policy* 15 (2002): 121–31.

7   See http://www.earthfirstjournal.org/section.php?id=1.
8   Donald Mann, president of Negative Population Growth, quoted in Garrett Hardin, *Living within Limits: Ecology, Economics, and Population Taboos* (New York: Oxford University Press, 1993), p. 206.
9   William McDonough and Michael Braungart, *Cradle to Cradle* (New York: North Point Press, 2002), p. 155.

Chapter 1   Sustainability and Time

1   See http://www.andykerr.net/KlamathBasin/TuleLakeRotarySpeech. html.
2   Edith Brown Weiss, *In Fairness to Future Generations* (Tokyo: The United Nations University, 1989), p. 38.
3   Al Gore, *Earth in the Balance* (Boston, MA: Houghton Mifflin, 1992), p. 170.
4   Ernest Partridge, "Why Care about the Future," pp. 203–20, in Ernest Partridge, ed., *Responsibilities to Future Generations* (Buffalo, NY: Prometheus Books, 1981), pp. 217–18.
5   Gifford Pinchot, *The Fight for Conservation* (New York: Doubleday, Page and Company, 1910), p. 79.
6   Aldo Leopold, *A Sand County Almanac, with Essays on Conservation from Round River* (New York: Ballantine Books, 1966), p. 240.
7   Ibid., p. 190; italics added.
8   Donella Meadows, Dennis Meadows, Jorgen Randers, and William Behrens, *The Limits to Growth* (New York: Universe Books, 1972), p. 158; italics added.
9   Ibid., p. 24.
10 Julian Simon, *The Ultimate Resource* (Princeton, NJ: Princeton University Press, 1981), pp. 42, 345.
11 Herman Kahn, William Brown, and Leon Martel, *The Next 200 Years: A Scenario for America and the World* (New York: William Morrow and Company: 1976), p. 1.
12 Donella Meadows, Jorgen Randers, and Dennis Meadows, *Limits to Growth: The 30-Year Update* (White River Junction, VT: Chelsea Green Publishing, 2004), p. 122.
13 International Union for the Conservation of Nature and Natural Resources, *World Conservation Strategy*, section 1, "Introduction: Living Resource Conservation for Sustainable Development"; accessed at data.iucn.org/dbtw-wpd/edocs/WCS-004.pdf. See also http://cmsdata.iucn.org/downloads/iucn _future_of_sustanability.pdf.
14 Lester R. Brown, *Building a Sustainable Society* (New York: W. W. Norton, 1981), pp. 8, 369.

15 World Commission on Environment and Development, *Our Common Future* (Oxford: Oxford University Press, 1987), p. 8.
16 John S. Dryzek, *The Politics of the Earth: Environmental Discourses* (Oxford: Oxford University Press, 1997), p. 123.
17 *Audubon Magazine*, July–August 1995, p. 6.
18 International Union for the Conservation of Nature and Natural Resources, *World Conservation Strategy*, section 1, "Introduction: Living Resource Conservation for Sustainable Development"; accessed at data.iucn.org/dbtw-wpd/edocs/WCS-004.pdf. See also http://cmsdata.iucn.org/downloads/iucn_future_of_sustanability.pdf. See David Brower, *Let the Mountains Talk, Let the Rivers Run: A Call to Those Who Would Save the Earth* (New York: HarperCollins, 1995), p. 1.
19 Edmund Burke, *Reflections on the Revolution in France* (Garden City, NY: Doubleday, 1961), p. 110.
20 Ibid., p. 108.
21 Wes Jackson, "Tackling the Oldest Environmental Problem," in Richard Heinberg and Daniel Lerch, eds, *The Post Carbon Reader* (Healdsburg, CA: Watershed Media, 2010), p. 129.
22 Michael Pollen, "Farmer in Chief," *New York Times Magazine*, October 12, 2009; accessible at http://michaelpollan.com/articles-archive/farmer-in-chief/.
23 Lester W. Milbrath was one of the first to employ this helpful visualization. See Lester W. Milbrath, *Envisioning a Sustainable Society* (Albany, NY: State University of New York Press, 1989), p. 2.
24 *Tao Te Ching*, trans. Stephen Mitchell (New York: Harper Collins, 1988), chapter 9 (np).
25 Aristotle, *The Ethics of Aristotle: The Nicomachean Ethics*, trans. J. A. K. Thomson (New York: Penguin Books, 1953), p. 173.
26 Confucius, *The Analects* (New York: Penguin, 1979), p. 134.
27 See http://www.ratical.org/many_worlds/6Nations/EoL/index.html#ToC; http://www.indigenouspeople.net/iroqcon.htm.
28 Jared Diamond, *Collapse: How Societies Choose to Fail or Succeed* (New York: Viking, 2005). See also Joseph A. Tainter, *The Collapse of Complex Societies* (Cambridge, UK: Cambridge University Press, 1988); Clive Ponting, *A Green History of the World: The Environment and Collapse of Great Civilizations* (New York: St Martin's Press, 1991).
29 Donald Rumsfeld, June 6, 2002, in address to a NATO meeting in Brussels; accessed at http://www.nato.int/docu/speech/2002/s020606g.htm.
30 R. Lewin, "In Ecology, Change Brings Stability," *Science* 234 (1986): 1071–3; Stuart Pimm, *The Balance of Nature?* (Chicago, IL: University of Chicago Press, 1991).

31   Quoted in Meadows, *Limits to Growth: The 30-Year Update*, p. 265.

Chapter 2   The Geography of Sustainability

1   Robert Gilman, quoted in *Choices for Sustainable Living* (Portland, OR: Northwest Earth Institute, 2010), p. 13.
2   WCED, *Our Common Future* (Oxford: Oxford University Press, 1987), p. 43.
3   Aron Sachs, "Upholding Human Rights and Environmental Justice," in Lester R. Brown et al., *State of the World 1996* (New York: W. W. Norton, 1996), p. 151.
4   Thomas C. Peterson, Peter A. Stott, and Stephanie Herring, eds, "Explaining Extreme Events of 2011 from a Climate Perspective," National Oceanographic and Atmospheric Administration, 2012; available at: http://www1.ncdc.noaa.gov/pub/data/cmb/bams-sotc/2011-peterson-et-al.pdf.
5   See Simon Dalby, *Security and Environmental Change* (Cambridge: Polity, 2009); and observe the concerns of the United Nations Refugee Agency at http://www.unhcr.org/pages/49c3646c10a.html.
6   Richard Douthwaite, "The International Response to Climate Change," in Richard Heinberg and Daniel Lerch, eds, *The Post Carbon Reader* (Healdsburg, CA: Watershed Media, 2010), p. 55.
7   Quoted in Peter Borrelli, "The Ecophilosophers," in Peter Borrelli, ed., *Crossroads: Environmental Priorities for the Future* (Washington, DC: Island Press, 1988), p. 69.
8   Aristotle, *The Politics and the Constitution of Athens* (Cambridge: Cambridge University Press, 1996), p. 33.
9   Paul R. Ehrlich, *The Population Bomb* (New York: Ballantine Books, 1968), p. 9.
10 Elinor Ostrom, *Governing the Commons: The Evolution of Institutions for Collective Action* (Cambridge: Cambridge University Press, 1990).
11 David Bollier, *Think Like a Commoner: A Short Introduction to the Life of the Commons* (Gabriola Island, BC: New Society Publisher, 2014), p. 176.
12 Karl Marx and Friedrich Engels, "Manifesto of the Communist Party," in Robert C. Tucker, ed., *The Marx-Engels Reader*, 2nd edn (New York: W. W. Norton, 1978), pp. 469–500.
13 See the KOF Index of Globalization, http://globalization.kof.ethz.ch/.
14 See the CIA *World Factbook*; accessible at https://www.cia.gov/library/publications/the-world-factbook/index.html.

15 Thomas L. Friedman, *The World Is Flat* (New York: Farrar, Straus and Giroux, 2006), p. 522.

16 See the Gini Index of distribution of family income in the CIA's *World Factbook*, at https://www.cia.gov/library/publications/the-world-factbook.

17 The CEOs of the 500 largest US corporations on average earned under US$2.5 million total compensation in 1990 and approximately US$40,000. See: http://www.forbes.com/2010/04/28/compensation-chief-executive-highest-paid-leadership-boss-10-ceo-compensation_land.html.

18 See https://www.census.gov/hhes/www/poverty/about/overview/; Marisol Bello, "Families in Extreme Poverty Double," *USA Today*, February 24, 2012, p. 3A.

19 Happy Planet Index 2.0, 8, http://www.happyplanetindex.org/public-data/files/happy-planet-index-2-0.pdf.

20 Richard Heinberg, *The End of Growth: Adapting to Our New Economic Reality* (Gabriola Island, BC, Canada: New Society Publishers, 2011), p. 225. See also Susan George, *Whose Crisis, Whose Future: Towards a Greener, Fairer, Richer World* (Cambridge: Polity, 2010), p. 83.

21 See http://www.dni.gov/nic/NIC_globaltrend2015.html.

22 See http://www.ifg.org/about.htm.

23 Barry Commoner, "How Poverty Breeds Overpopulation (and Not the Other Way Around)," *Ramparts* 13(10) (1974): 58–9, excerpted in Carolyn Merchant, ed., *Ecology: Key Concepts in Critical Theory* (Atlantic Highlands, NJ: Humanities Press, 1994), pp. 88–95; Barry Commoner, *Making Peace with the Planet* (New York: Pantheon Books, [1975] 1990).

24 Peter Wenz, "Does Environmentalism Promote Injustice for the Poor," in Ronald Sandler and Phaedra Pezzulo, eds, *Environmental Justice and Environmentalism: The Social Justice Challenge to the Environmental Movement* (Cambridge, MA: MIT Press, 2007), pp. 57–83.

25 Vandana Shiva, *Earth Democracy: Justice, Sustainability, and Peace* (Cambridge, MA: South End Press, 2005), p. 61.

26 Quoted in Stephan Schmidheiny, *Changing Course: A Global Business Perspective on Development and the Environment* (Cambridge, MA: MIT Press, 1992), p. 135.

27 Richard Wilkinson and Kate Pickett, *The Spirit Level: Why Greater Equality Makes Societies Stronger* (New York: Bloomsbury Press, 2009).

28 Frederick Solt, "Economic Inequality and Democratic Political Engagement," *American Journal of Political Science* 52 (January 2008): 48–60.

29   Mariano Torras and James Boyce, "Income, Inequality, and Pollution: A Reassessment of the Environmental Kuznets Curve," *Ecological Economics* 25 (1998): 147–60; J. Boyce, A. Klemer, P. Templet, and C. Willis, "Power Distribution, the Environment, and Public Health: A State-Level Analysis," *Ecological Economics* 29 (1999): 127–40.

30   WCED, *Our Common Future*, p. 6.

31   See Robert Putnam, "E Pluribus Unum: Diversity and Community in the Twenty-first Century: The 2006 Johan Skytte Prize Lecture," *Scandinavian Political Studies* 30 (June 2007): 137–74.

32   Julian Agyeman, Robert Bullard, and Bob Evans, "Towards Just Sustainabilities: Perspectives and Possibilities," in Julian Agyeman, Robert Bullard, and Bob Evans, eds, *Just Sustainabilities: Development in an Unequal World* (Cambridge, MA: MIT Press, 2003), p. 325.

33   Andreas Bergh and Therese Nilsson, "Good for Living? On the Relationship between Globalization and Life Expectancy," *World Development* 38 (2010): 1191–1203.

34   R. Layard, G. Mayraz, and S. Nickell, "The Marginal Utility of Income," *Journal of Public Economics* 92 (2008): 1846–57; Angus Deaton, "Income, Health and Wellbeing around the World: Evidence from the Gallup World Poll," *Journal of Economic Perspectives* 22 (2008): 53–77, at http://www.ncbi. nlm.nih.gov/pmc/articles/PMC2680297/; Wilkinson and Pickett, *The Spirit Level*, pp. 31–45.

35   Michael Shermer, *The Mind of the Market* (New York: Holt, 2008), 147–8. Wilkinson and Pickett, *The Spirit Level*, pp. 31–45.

36   See Agyeman, Bullard, and Evans, "Towards Just Sustainabilities."

37   Ulrich Beck, *Risk Society: Towards a New Modernity* (Thousand Oaks, CA: Sage Publications, 1986), p. 35.

38   See United Church of Christ Commission for Racial Justice, *Toxic Wastes and Race in the United States* (New York: United Church of Christ Commission for Racial Justice, 1987). Robert Bullard, *Dumping in Dixie: Race, Class and Environmental Quality* (Boulder, CO: Westview Press, 1990). Robert Bullard, ed., *Unequal Protection: Environmental Justice and Communities of Color* (New York: Random House, 1994).

39   Van Jones, "Bridging the Green Divide," in *Choices for Sustainable Living* (Portland, OR: Northwest Earth Institute, 2010), p. 98.

40   Kristin Shrader-Frechette, *Environmental Justice: Creating Equality, Reclaiming Democracy* (Oxford: Oxford University Press, 2002), p. 3.

41   Quoted from *Adlai Today*, http://www.adlaitoday.org/ideas/connect_sub2_engage.html.
42   Quoted in Marie-Claire Cordonier Segger and C. G. Weeramantry, eds, *Sustainable Justice* (The Netherlands: Martinus Nijhoff, 2005), p. 1.
43   The World Conservation Union, United Nations Environment Programme, World Wide Fund for Nature, *Caring for the Earth: A Strategy for Sustainable Living* (Gland, Switzerland, 1991), p. 5.
44   See Garrett Hardin, *Living within Limits: Ecology, Economics, and Population Taboos* (New York: Oxford University Press, 1993), p. 278.
45   Wendell Berry, *Sex, Economy, Freedom and Community* (New York: Pantheon Books, 1993), p. 19.
46   Ibid., p. 24.
47   Susan M. Koger and Deborah Du Nann Winter, *The Psychology of Environmental Problems*, 3rd edn (New York: Taylor and Francis, 2010), pp. 111, 126, 130, 135, 317–18.
48   Tim Jackson, *Prosperity without Growth: Economics for a Finite Planet* (London: Earthscan, 2009), pp. 144–5.
49   See Kirkpatrick Sale, *Dwellers in the Land: The Bioregional Vision* (New York: Random House, 1985).
50   Katharine Mieszkowski, "Plastic Bags are Killing Us," in *Choices for Sustainable Living* (Portland, OR: Northwest Earth Institute, 2010), p. 49.
51   William McDonough and Michael Braungart, *Cradle to Cradle* (New York: North Point Press, 2002).
52   Aristotle, *The Politics* (Cambridge: Cambridge University Press, 1996), p. 108.
53   See http://www.dni.gov/nic/NIC_globaltrend2015.html.
54   See http://www.ifg.org/about.htm.

Chapter 3   Ecological Resilience and Environmental Health

1   Camilo Mora, Derek Tittensor, Sina Adi, Alastair Simpson, and Boris Worm, "How Many Species Are There on Earth and in the Ocean?" *PLoS Biology* 9(8) (2011): e1001127. doi:10.1371/journal.pbio.1001127.
2   International Union for the Conservation of Nature's Red List, at   http://www.iucn.org/about/work/programmes/species/red_list/about_the_red_list/.
3   Henry Fountain, "Too Many Small Fish are Caught, Report Says," *New York Times*, April 2, 2012, p. A9.

4   "Extinction Rate across the Globe Reaches Historical Propor-
     tions," *ScienceDaily*, January 10, 2002; accessed at http://
     www.sciencedaily.com/releases/2002/01/020109074801.htm.
5   *Ecosystems and Human Well-Being, Synthesis: A Report of the
     Millennium Ecosystem Assessment* (Washington, DC: Island
     Press, 2005), p. 36; accessed at http://www.millenniumasse
     ssment.org/en/index.aspx.
6   Michael Soulé, "Thresholds for Survival: Maintaining Fitness and
     Evolutionary Potential," in Michael Soulé and Bruce Wilcox, eds,
     Conservation Biology: An Evolutionary-Ecological Perspective
     (Sunderland, MA: Sinauer Associates, 1980), pp. 166, 168.
7   Arun Agrawal and Kent Redford, "Conservation and Displace-
     ment: An Overview," *Conservation and Society* 7 (2009):
     1–10. Mark Dowie, *Conservation Refugees: The Hundred-
     Year Conflict between Global Conservation and Native Peoples*
     (Boston, MA: MIT Press, 2009).
8   Lester Brown, *World on the Edge: How to Prevent Environ-
     mental and Economic Collapse* (New York: Earth Policy Insti-
     tute, 2011), p. 146.
9   R. P. Singh, "Causes and Consequences of the Greenhouse
     Gases," *Bulletin of Pure and Applied Sciences, Section F: –
     Geology* 25(1–2) (January 2006).
10   Edward O. Wilson, *Biophilia* (Cambridge, MA: Harvard Uni-
     versity Press, 1984), p. 121.
11   Stephanie Mills, "Peak Nature," in Richard Heinberg and
     Daniel Lerch, eds, *The Post Carbon Reader* (Healdsburg, CA:
     Watershed Media, 2010), p. 103.
12   Paul Ehrlich, "Population Diversity and the Future of Ecosys-
     tems," *Science* 254 (1991): 175.
13   Holly Jones and Oswald Schmitz, "Rapid Recovery of Damaged
     Ecosystems," *PLoS One* 4/5: e5653. *PLoS One* 4/5: e5653, at
     http://www.plosone.org/article/info:doi/10.1371/journal.pone.
     0005653.
14   Plato, *The Collected Dialogues* (Princeton, NJ: Princeton Uni-
     versity Press, 1989), pp. 1216–17.
15   George Perkins Marsh, *Man and Nature: Physical Geography
     as Modified by Human Action* (New York: C. Scribner's Sons,
     1869), pp. 9, 95, 108, 186.
16   Paul Ehrlich and Anne Ehrlich, *Extinction: The Causes and
     Consequences of the Disappearance of Species* (New York:
     Random House, 1981).
17   "Economic Value of Insect Pollination Worldwide Estimated
     at U.S. $217 Billion," *Science Daily*, September 15, 2008;
     available at http://www.sciencedaily.com/releases/2008/09/08
     0915122725.htm.

18 Robert Costanza et al., "The Value of the World's Ecosystem Services and Natural Capital," *Nature* 387 (May 1997): 253–60.

19 Gretchen Daily, ed., *Nature's Services: Societal Dependence on Natural Ecosystems* (Washington, DC: Island Press, 1997), p. 365.

20 Pushpamm Kumar, ed., *The Economics of Ecosystems and Biodiversity: Ecological and Economic Foundations* (London: Earthscan, 2010).

21 Gretchen Daily et al., "The Value of Nature and the Nature of Value," *Science* 289 (July 2000): 395–6.

22 See Information Center for the Environment, at http://ice.ucdavis.edu/node/133.

23 Daily et al., "The Value of Nature and the Nature of Value," pp. 395–6.

24 See Michael J. Sandel, *What Money Can't Buy: The Moral Limits of Markets* (New York: Farrar, Straus, and Giroux, 2012).

25 Bill Mollison, *Permaculture: A Practical Guide for a Sustainable Future* (Washington, DC: Island Press, 1990), p. 507.

26 Wilson, *Biophilia*.

27 Stephen Kellert, *Building for Life: Designing and Understanding the Human–Nature Connection* (Washington, DC: Island Press, 2006), p. 4.

28 See Warren Hern's "Has the Human Species Become a Cancer on the Planet? A Theoretical View of Population Growth as a Sign of Pathology," available at http://www.drhern.com/pdfs/humancancerplanet.pdf.

29 Kirk Smith, Carlos Corvalan, and Tord Kjellstroem, "How Much Global Ill Health is Attributable to Environmental Factors?" *Epidemiology* 10 (September 1999): 573–84. Kathryn H. Jacobsen, *Introduction to Global Heath* (Boston, MA: Jones and Bartlett, 2008), pp. 230–1.

30 Theo Colburn, Dianne Dumanoski, and John Peter Meyers, *Our Stolen Future* (New York: Plume, 1997); Joe Thornton, *Pandora's Poison* (Cambridge, MA: MIT Press, 2000); Susan M. Koger and Deborah Du Nann Winter, *The Psychology of Environmental Problems*, 3rd edn (New York: Taylor and Francis, 2010), pp. 166–84. Liz Szabo, "Pregnant Women Rife with Chemicals," *USA Today*, January 14, 2011, 3A.

Chapter 4   Technological Solutions and Their Problems

1 Amory Lovins and Rocky Mountain Institute, *Reinventing Fire: Bold Business Solutions for the New Energy Era* (White River Junction, VT: Chelsea Green Publishing, 2011), pp. 82–3. C. Kibert, J. Sendzimir and B. Guy, "Construction

Ecology and Metabolism: Natural System Analogues for a Sustainable Built Environment," *Construction Management & Economics* 18 (2000): 903–16. L. Hunter Lovins and Boyd Cohen, *Climate Capitalism: Capitalism in the Age of Climate Change* (New York: Hill and Wang, 2011), p. 94.

2 See *The World Economic and Social Survey*, and accompanying press release, at http://www.un.org/en/development/desa/

3 Amory Lovins and Rocky Mountain Institute, *Reinventing Fire*, p. 92.

4 Paul R. Ehrlich, John P. Holdren, and Barry Commoner, "A Bulletin Dialogue: Critique and Response," *Bulletin of the Atomic Scientists* (May 1972): 16–27, 42–56.

5 Marian R. Chertow, "The IPAT Equation and Its Variants: Changing Views of Technology and Environmental Impact," *Journal of Industrial Ecology* 4(4) (fall 2000): 13–29.

6 Jared Diamond, *Collapse: How Societies Choose to Fail or Succeed* (New York: Viking, 2005).

7 Leslie Paul Thiele, *Indra's Net and the Midas Touch: Living Sustainably in a Connected World* (Cambridge, MA: MIT Press, 2011).

8 See Hilary F. French, "Learning from the Ozone Experience," in Lester R. Brown et al., *State of the World 1997* (New York: W. W. Norton, 1997), pp. 151–71; "Can We Save Our Skins?" *Friends of the Earth* (July/August 1996): 8–11.

9 Wendell Berry, "The Way of Ignorance," in Bill Vitek and Wes Jackson, eds, *The Virtues of Ignorance: Complexity, Sustainability, and the Limits of Knowledge* (Lexington, KY: University of Kentucky Press, 2008), p. 37.

10 See the Rodale Institute's white paper: http://rodaleinstitute. org/assets/RegenOrgAgricultureAndClimateChange_20140 418.pdf.

11 Albert Bates and Toby Hemenway, "From Agriculture to Permaculture," in *2010 State of the World: Transforming Cultures* (Washington, DC: Worldwatch Institute, 2010), pp. 50–1; Rodale Institute, "Organic Farming May Be the Best Route to Global Food Security," in *Choices for Sustainable Living* (Portland, OR: Northwest Earth Institute, 2010), pp. 67–8. L. Hunter Lovins and Boyd Cohen, *Climate Capitalism: Capitalism in the Age of Climate Change* (New York: Hill and Wang, 2011), pp. 189, 194–5.

12 Verena Seufert, Navin Ramankutty, and Jonathan A. Foley, "Comparing the Yields of Organic and Conventional Agriculture," *Nature* 485 (May 10, 2012): 229–32.

13 See http://nobelprize.org/nobel_prizes/peace/articles/borlaug/ borlaug-lecture.pdf, p. 21.

14 Union of Concerned Scientists, "Impacts of Genetically Engineered Crops on Pesticide Use in the United States: The First Thirteen Years," *Catalyst* (spring 2010): 5.

15 Bill Mollison, *Permaculture: A Practical Guide for a Sustainable Future* (Washington, DC: Island Press, 1990), p. 34.

16 See http://www.gdrc.org/u-gov/precaution-3.html.

17 World Commission on the Ethics of Scientific Knowledge and Technology, *The Precautionary Principle* (Paris: UNESCO, 2005), p. 16; accessed at http://portal.unesco.org/shs/en/ev.php -URL_ID=7694&URL_DO=DO_TOPIC&URL_SECTION =201.html.

18 Kai N. Lee, *Compass and Gyroscope: Integrating Science and Politics for the Environment* (Washington, DC: Island Press, 1993), pp. 8–9.

19 Ralf Yorque, Brian Walker, C. S. Holling, Lance Gunderson, Carl Folke, Stephen Carpenter, and William Brock, "Toward an Integrative Synthesis," in Lance H. Gunderson and C. S. Holling, eds, *Panarchy: Understanding Transformations in Human and Natural Systems* (Washington, DC: Island Press, 2002), p. 438. See also Lee, *Compass and Gyroscope*, p. 138.

20 C. Walters and C. S. Holling, "Large-scale Management Experiments and Learning by Doing," *Ecology* 71 (1990): 2060–8. R. Gregory, D. Ohlson, and J. Arvai, "Deconstructing Adaptive Management: Criteria for Applications to Environmental Management," *Ecological Applications* 16 (2006): 2411–25.

21 Lee, *Compass and Gyroscope*, pp. 11, 53.

22 Wendell Berry, "Solving for Pattern," in *The Gift of Good Land* (San Francisco, CA: North Point Press, 1981), pp. 137–8.

23 Union of Concerned Scientists, *Earthwise* 13(3) (summer 2011): 3.

24 See http://www.cdc.gov/drugresistance/ and Gardiner Harris, "Prescription Rule Will Cut Use of Antibiotics in Livestock, F.D.A. Says," *New York Times*, April 12, 2012, p. A17.

25 See http://www.news.cornell.edu/releases/aug97/livestock.hrs. html.

26 Berry, "Solving for Pattern," p. 143.

27 Wendell Berry, *What are People For?* (San Francisco, CA: North Point Press, 1990), p. 166.

28 Peter M. Allen, "Complexity in Socio-economic Systems" in L. Douglas Kiel, ed., *Knowledge Management, Organizational Intelligence and Learning, and Complexity*, Vol. 2 (Oxford: Eolss Publishers/UNESCO, 2009), pp. 118–37, p. 133.

Chapter 5   The Political and Legal Challenge

1   Susan George, *Whose Crisis, Whose Future: Towards a Greener, Fairer, Richer World* (Cambridge: Polity, 2010), p. 2.
2   See https://www.whitehouse.gov/sites/default/files/omb/inforeg/ 2013_cb/draft_2013_cost_benefit_report.pdf.
3   Richard H. Thaler and Cass R. Sunstein, *Nudge: Improving Decisions about Health, Wealth, and Happiness* (New Haven, CT: Yale University Press, 2008).
4   "Free Exchange: Nudge Nudge, Think Think," *The Economist*, March 24, 2012, p. 78.
5   Gifford Pinchot, *The Fight for Conservation* (New York: Doubleday, Page and Company, 1910), pp. 26, 46, 81–2.
6   WCED, *Our Common Future* (Oxford: Oxford University Press, 1987), p. 65.
7   Larry Lohmann, "Whose Common Future," *The Ecologist* 20/3 (May/June 1990): 82–4. On the question of the potentially pernicious uses of the terms "environmental security" and "ecological security," see also Lothar Brock, "Security Through Defending the Environment: An Illusion?," in Elise Boulding, ed., *New Agendas for Peace Research: Conflict and Security Reexamined* (Boulder, CO: Lynne Rienner, 1992), pp. 79–102.
8   George, *Whose Crisis, Whose Future*, p. 57.
9   Donella H. Meadows, *Thinking in Systems* (White River Junction, VT: Chelsea Green Publishing, 2008), p. 173.
10   See http://www.epi.yale.edu/.
11   See http://www.ssfindex.com/.
12   See http://www.happyplanetindex.org/. See also the Calvert-Henderson Quality of Life Indicators, at http://www.calvert -henderson.com/; the Genuine Progress Indicator produced by Redefining Progress, at http://www.solar783.com/solar783/ gpi1999.pdf; and the Human Development Reports compiled by the United Nations Development Program, at http://hdr. undp.org/en/statistics/.
13   See http://www.footprintnetwork.org.
14   WWF Living Planet Report, 2010, at http://wwf.panda.org/ about_our_earth/all_publications/living_planet_report/.
15   See http://www.legislation.govt.nz/act/public/1991/0069/latest/ DLM231905.html#DLM231905.
16   See http://www.sejm.gov.pl/prawo/konst/angielski/kon1.htm.
17   See http://www.iuscomp.org/gla/statutes/GG.htm#20a.
18   See http://venezuelanalysis.com/constitution/title/3.
19   See   http://www.legifrance.gouv.fr/html/constitution/const03. htm.

20 See Joerg Chet Tremmel, "Establishing Intergenerational Justice in National Constitutions," in Joerg Chet Tremmel, ed., *Handbook of Intergenerational Justice*, Cheltanham: Edward Elgar, 2006), pp. 187–214.

21 See http://www.unep.org/Documents.multilingual/Default.asp? DocumentID=78&ArticleID=1163.

22 See http://www.cites.org/eng/disc/what.shtml.

23 See http://untreaty.un.org/cod/avl/ha/bcctmhwd/bcctmhwd. html.

24 See http://www.icj-cij.org/homepage/index.php?lang=en.

25 Justin Gillis, "Climate Accord Is a Healing Step, if Not a Cure," *New York Times*, December 12, 2015, accessed at http://www.nytimes.com/2015/12/13/science/earth/climate-accord-is-a-healing-step-if-not-a-cure.html?emc=edit_th_2015 1213&nl=todaysheadlines&nlid=71023177.

26 See the National Oceanic and Atmospheric Administration report at http://www.ncdc.noaa.gov/sotc/summary-info/global/ 201512.

27 See Thomas F. Homer-Dixon's works in this area, including "Environmental Scarcities and Violent Conflict: Evidence from Cases," *International Security* 19(1) (1994): 5–40; *Environment, Scarcity, and Violence* (Princeton, NJ: Princeton University Press, 2001); and, with a broad historical sweep, *The Upside of Down: Catastrophe, Creativity, and the Renewal of Civilization* (Toronto: Knopf, 2006).

28 Norman Myers, *Ultimate Security: The Environmental Basis of Political Stability* (Washington, DC: Island Press, 1993), pp. 12–13, 231.

29 Lester Brown, *Plan B 4.0: Mobilizing to Save Civilization* (New York: Norton, 2009), p. 6.

30 See http://www.indexmundi.com/commodities/?commodity -food-price-index&months=240.

31 See, for instance, http://www.guardian.co.uk/world/2011/ jan/07/algeria-riots-food-prices.

32 Brown, *Plan B 4.0*, pp. 186–7.

33 W. Neil Adger and Andrew Jordan, eds, "Preface" to *Governing Sustainability* (Cambridge: Cambridge University Press, 2009), p. xvii.

34 Brown, *World on the Edge: How to Prevent Environmental and Economic Collapse* (New York: Earth Policy Institute, 2011), pp. 200–2.

35 Quoted at http://www.state.gov/s/partnerships/ubuntu/index.htm.

36 Mollison, *Permaculture: A Practical Guide for a Sustainable Future* (Washington, DC: Island Press), 1990, p. 506.

Chapter 6   Sustaining Economies

1   Amy E. Knaup, "Survival and Longevity in the Business Employment Dynamics Data," *Monthly Labor Review* (May 2005): 51(50–6); available at http://smallbiztrends.com/2005/07/business-failure-rates-highest-in.html.

2   Ray Anderson, with Robin White, *Confessions of a Radical Industrialist: Profits, People, Purpose – Doing Business by Respecting the Earth* (St Martins, 2009); Daniel C. Esty and Andrew Winston, *Green to Gold: How Smart Companies Use Environmental Strategy to Innovate, Create Value, and Build Competitive Advantage* (Hoboken, NJ: Wiley, 2009). L. Hunter Lovins and Boyd Cohen, *Climate Capitalism: Capitalism in the Age of Climate Change* (New York: Hill and Wang, 2011).

3   "Schumpeter: Good Business; Nice Beaches," *The Economist*, May 19, 2012, p. 76.

4   Deirdre McCloskey, *The Bourgeois Virtues: Ethics for an Age of Commerce* (Chicago, IL: University of Chicago Press, 2006), p. 19.

5   Herman Daly, *Beyond Growth: The Economics of Sustainable Development* (Boston, MA: Beacon Press, 1996).

6   Center for Sustainable Systems, University of Michigan, "U.S. Material Use," accessed at www.css.snre.umich.edu/css_doc/CSS05-18.pdf.

7   The earth has a mass of 6×10^{24} kilograms. Doubling 6.5×10^9 metric tons of raw material consumed every 7.2 years gives us a total of 6×10^{24} kilograms consumed in about 300 years.

8   Tim Jackson, *Prosperity without Growth: Economics for a Finite Planet* (London: Earthscan), 2009, p. 88.

9   Paul Hawken, Amory Lovins, and L. Hunter Lovins, *Natural Capitalism* (New York: Little, Brown, 1999), pp. 8, 14.

10   Quoted in Mike Nickerson, *Life, Money and Illusion: Living on Earth as if We Want to Stay* (Lanark, Ontario: Seven Generations Publishing, 2006), p. 158.

11   William Ophuls and A. Stephen Boyan, Jr, *Ecology and the Politics of Scarcity Revisited: The Unraveling of the American Dream* (New York: W. H. Freeman, 1992), pp. 237–9.

12   Herman E. Daly, "The Steady-State Economy: Toward a Political Economy of Biophysical Equilibrium and Moral Growth," in *Toward a Steady-State Economy* (San Francisco, CA: W. H. Freeman, 1973), p. 167.

13   Heinberg, *The End of Growth: Adapting to Our New Economic Reality* (Gabriola Island, BC, Canada: New Society Publishers, 2011).

14 John Stuart Mill, *Principles of Political Economy*, in V. Bladen and J. Robson, eds, *The Collected Works of John Stuart Mill* (Toronto: University of Toronto Press, 1965), p. 754.

15 The World Conservation Union, United Nations Environment Programme, World Wide Fund for Nature, *Caring for the Earth: A Strategy for Sustainable Living* (Gland, Switzerland, 1991), p. 10.

16 The Earth Charter is available at http://www.earthcharterin action.org/content/pages/Read-the-Charter.html.

17 David Selby, "The Firm and Shaky Ground of Education for Sustainable Development," *Journal of Geography in Higher Education* 30(2): 351–65. Magnus Boström, "A Missing Pillar? Challenges in Theorizing and Practicing Social Sustainability," *Sustainability Science, Practice, and Policy* 8(1): 3–14 (2012).

18 David. W. Orr, *The Nature of Design: Ecology, Culture, and Human Intention* (Oxford: Oxford University Press, 2002), p. 173.

19 Donald Worster, "The Shaky Ground of Sustainability," in George Sessions, ed., *Deep Ecology for the 21st Century* (Boston, MA: Shambhala, 1995), pp. 424, 417–18.

20 Johan Rockström et al., "A Safe Operating Space for Humanity," *Nature* 461 (September 2009), pp. 472–5. Will Steffen et al., "Planetary Boundaries: Guiding Human Development on a Changing Planet," *Science* 347(6223) (February 13, 2015): DOI: 10.1126/science.1259855

21 See http://www.who.int/features/factfiles/water/water_facts/en/index.html.

22 See http://thewaterproject.org/water_scarcity_2.asp.

23 See http://www.unesco.org/water/wwap/facts_figures/food_supply.shtml; http://www.news.cornell.edu/releases/aug97/live-stock.hrs.html.

24 Juliet Schor, *True Wealth* (New York: Penguin, 2011), p. 64.

25 See www.waterfootprint.org.

26 Sandra Postel, "Water," in Richard Heinberg and Daniel Lerch, eds, *The Post Carbon Reader* (Healdsburg, CA: Watershed Media, 2010), p. 87.

27 See http://www.unesco.org/water/wwap/facts_figures/basic_needs.shtml.

28 Laurence C. Smith, *The World in 2050* (New York: Penguin, 2011), pp. 100–3.

29 See http://www.guardian.co.uk/environment/2010/feb/18/worlds-top-firms-environmental-damage.

30 David Bollier, *Think like a Commoner: A Short Introduction to the Life of the Commons* (Gabriola Island, BC: New Society Publisher, 2014), p. 109.

31   Paul Hawken, *The Ecology of Commerce: A Declaration of Sustainability* (New York: HarperCollins, 1993), p. 13.
32   http://www.worldenergyoutlook.org/resources/energysubsidies/. See also: L. Hunter Lovins and Boyd Cohen, *Climate Capitalism: Capitalism in the Age of Climate Change* (New York: Hill and Wang, 2011), pp. 19, 278; Frances Moore Lappé, *Getting a Grip 2* (Cambridge, MA: Small Planet Media, 2010), pp. 80, 84; Nickerson, *Life, Money and Illusion*, p. 264.
33   John Antle and Gregg Heidebrink, "Environment and Development: Theory and International Evidence," *Economic Development and Cultural Change* 43 (April 1995): 603–25.
34   For a broad perspective on the "race to the bottom" debate, see Jennifer Clapp, "What the Pollution Havens Debate Overlooks," *Global Environmental Politics* 2(2) (May 2002): 11–19.
35   "Free Exchange: Decline and Small," *The Economist*, March 3, 2012, p. 87.

Chapter 7   Culture and Change

1   See Tim Hartford, *Adapt: Why Success Always Starts with Failure* (New York: Farrar, Straus and Giroux, 2011), p. 160.
2   Union of Concerned Scientists, "Smarter Ways to Keep the Planet Cool," *Catalyst* (spring 2012): 7–8.
3   Smith, *The World in 2050* (New York: Penguin, 2011), p. 99.
4   Alasdair MacIntyre, *After Virtue: A Study in Moral Theory* (Notre Dame, IN: University of Notre Dame Press, 1981), p. 201.
5   Thomas Berry, *The Dream of the Earth* (San Francisco, CA: Sierra Club Books, 1988), p. 123.
6   Aldo Leopold, *A Sand County Almanac, with Essays on Conservation from Round River* (New York: Ballantine Books, 1966), p. 239.
7   Ibid.
8   Martin Luther King Jr, "Letter from Birmingham City Jail," in James Melvin Washington, ed., *A Testament of Hope: The Essential Writings and Speeches of Martin Luther King, Jr* (San Francisco, CA: Harper, 1991), p. 290.
9   Confucius, *The Analects* (New York: Penguin, 1979), p. 135.
10   Matthew 7:12, also Luke 6:31, New Jerusalem Bible (New York: Doubleday 1985), p. 158.
11   John Rawls, *A Theory of Justice*, rev. edn (Cambridge, MA: Harvard University Press, 1999), pp. 225, 295. Rawls extends his moral framework to encompass intergenerational affairs in *Political Liberalism* (New York: Columbia University Press, 2005), p. 274.

12 Lynn White, Jr, "The Historical Roots of Our Ecologic Crisis," *Science* 155 (1967): 1203–7.

13 See, for example, the *Encyclopedia of Religion and Nature* (London: Continuum, 2008).

14 Leopold, *A Sand County Almanac*, p. 263.

15 Ibid., p. 262.

16 Rachel Carson, "The Real World Around Us," in Linda Lear, ed., *Lost Woods* (Boston, MA: Beacon Press, 1998), pp. 148–63.

17 Rachel Carson, *The Sense of Wonder* (New York: Harper, 1998), p. 100.

18 Esther Sternberg, *Healing Spaces: The Science of Place and Well-Being* (Cambridge, MA: Belknap Press, 2009).

19 Timothy Beatley, *Biophilic Cities: Integrating Nature into Urban Design and Planning* (Washington, DC: Island Press, 2011), pp. 1–16.

20 David Orr, *The Nature of Design: Ecology, Culture, and Human Intention* (Oxford: Oxford University Press, 2002), p. 201.

21 Susan Linn, "Commercialism in Children's Lives," in *2010 State of the World: Transforming Cultures* (Washington, DC: Worldwatch Institute, 2010), pp. 63, 65.

22 Ingrid Pramling Samuelsson and Yoshie Kaga, "Early Childhood Education to Transform Cultures for Sustainability," *2010 State of the World: Transforming Cultures* (Washington, DC: Worldwatch Institute, 2010), p. 58.

23 Susan Linn, *Consuming Kids: The Hostile Takeover of Childhood* (New York: New Press, 2004); Hawken, *The Ecology of Commerce*, p. 214.

24 Richard Louv, *Last Child in the Woods: Saving Our Children from Nature-Deficit Disorder* (Chapel Hill, NC: Algonquin Books, 2005).

25 Wilson, *Biophilia*, p. 139. See also Stephen Kellert and Edward O. Wilson, eds, *Biophilia Hypothesis* (Washington, DC: Island Press, 1993).

26 See http://www.ulsf.org/about.html.

27 See http://www.iisd.org/educate/declarat/coper.htm.

28 Donella Meadows, "Whole Earth Models and Systems," *The Coevolution Quarterly* 34 (summer 1982): 98–108.

29 See http://www.unep.org/.

30 Arjen Wals, *Review of Contexts and Structures for Education for Sustainable Development: Learning for a Sustainable World* (Paris: UNESCO, 2009), p. 49; Marco Rieckmann, "Future-Oriented Higher Education: Which Key Competencies Should Be Fostered through University Teaching and Learning?," *Futures* 44(2) (March 2012): 127–35.

31   Michael Bonnett, "Issues for Environmental Education," *Journal of Philosophy of Education* 37(4) (2003): 696 (691–705).

32   See Matthias Finger, "From Knowledge to Action? Exploring the Relationships between Environmental Experiences, Learning, and Behavior," *Journal of Social Issues* 50(3) (fall 1994): 147, 157–8. Doug Mackenzie-Mohr and William Smith, *Fostering Sustainable Behavior* (Gabriola Island, BC: New Society Publishers, 1999), pp. 9–11.

33   See Willett Kempton, James Boster, and Jennifer Hartley, *Environmental Values in American Culture* (Cambridge, MA: MIT Press, 1995), pp. 211–26.

34   Archibald P. Sia, Harold R. Hungerford, and Audrey N. Tomera, "Selected Predictors of Responsible Environmental Behavior: An Analysis," *Journal of Environmental Education* 17(2) (winter 1985/1986): 31–40. Jody M. Hines, Harold R. Hungerford, and Audrey N. Tomera, "Analysis and Synthesis of Research on Responsible Environmental Behavior: A Meta-Analysis," *Journal of Environmental Education* 18(2) (winter 1986–7): 1–8. Harold R. Hungerford and Trudi L. Volk, "Changing Learner Behavior through Environmental Education," *Journal of Environmental Education* 21(3) (spring 1990): 8–21. Peter Martin, "A WWF View of Education and the Role of NGOs," in John Huckle and Stephen Sterling, *Education for Sustainability* (London: Earthscan, 1996), pp. 40–51.

35   David W. Orr, *Earth in Mind: On Education, Environment, and the Human Prospect* (Washington, DC: Island Press, 1994), p. 8.

36   A. Bowers, *Educating for an Ecologically Sustainable Culture* (Albany, NY: State University of New York Press, 1995), p. 199.

37   Gregory A. Smith, *Education and the Environment: Learning to Live with Limits* (Albany, NY: State University of New York Press, 1992), p. 94.

38   See http://www.scientificamerican.com/article.cfm?id=what-me-care.

39   William Ryerson, "Population," in Richard Heinberg and Daniel Lerch, eds, *The Post Carbon Reader* (Healdsburg, CA: Watershed Media, 2010), p. 167.

40   Sara Parkin, *The Positive Deviant: Sustainability Leadership in a Perverse World* (London: Earthscan, 2010), p. 4.

41   Richard Norgaard, "Coevolution of Economy, Society and Environment," in Paul Ekins and Manfred Max-Neef, *Real-life Economics* (London: Routledge, 1992), pp. 781–2, 86. Leslie Paul Thiele, *Environmentalism for a New Millennium: The Challenge of Coevolution* (New York: Oxford University Press, 1999).

Conclusion

1 Richard Heinberg, *Powerdown: Options and Actions for a Post-Carbon World* (Gabriola Island, BC: New Society Publishers, 2004), p. 23.

2 A message on a popular sign, cited in Joseph A. Tainter, *The Collapse of Complex Societies* (Cambridge, UK: Cambridge University Press, 1988), pp. 193, 216.

3 Tim Jackson, *Prosperity without Growth: Economics for a Finite Planet* (London: Earthscan, 2009), pp. 143–56.

Select Bibliography

Adger, W. Neil and Andrew Jordan (eds). 2009. *Governing Sustainability*. Cambridge: Cambridge University Press.

Agrawal, Arun and Kent Redford. 2009. "Conservation and Displacement: An Overview," *Conservation and Society* 7: 1–10.

Agyeman, Julian, Robert Bullard, and Bob Evans. 2003. "Towards Just Sustainabilities: Perspectives and Possibilities," in Julian Agyeman, Robert Bullard, and Bob Evans, eds, *Just Sustainabilities: Development in an Unequal World*, Cambridge, MA: MIT Press, pp. 323–32.

Allen, Peter M. 2009. "Complexity in Socio-economic Systems," in L. Douglas Kiel, ed., *Knowledge Management, Organizational Intelligence and Learning, and Complexity*, Vol. 2: 118–37. Oxford: Eolss Publishers/UNESCO.

Anderson, Ray, with Robin White. 2009. *Confessions of a Radical Industrialist: Profits, People, Purpose – Doing Business by Respecting the Earth*. New York: St Martin's Press.

Antle, John and Gregg Heidebrink. April 1995. "Environment and Development: Theory and International Evidence," *Economic Development and Cultural Change* 43: 603–25.

Aristotle. 1953. *The Ethics of Aristotle: The Nicomachean Ethics*, trans. J. A. K. Thomson. New York: Penguin Books.

Aristotle. 1996. *The Politics and the Constitution of Athens*. Cambridge: Cambridge University Press.

Beatley, Timothy. 2011. *Biophilic Cities: Integrating Nature into Urban Design and Planning*. Washington, DC: Island Press.

Beck, Ulrich. 1986. *Risk Society: Towards a New Modernity*. Thousand Oaks, CA: Sage Publications.

Bello, Marisol. 2012. "Families in Extreme Poverty Double," *USA Today*, February 24, p. 3A.

Benbrook, Charles. November 2009. *Impacts of Genetically Engineered Crops on Pesticide Use in the United States: The First Thirteen Years*, accessed at: http://www.organic-center.org/reportfiles/13Years20091126_FullReport.pdf.

Bergh, Andreas and Therese Nilsson. 2010. "Good for Living? On the Relationship between Globalization and Life Expectancy," *World Development* 38: 1191–1203.

Berry, Thomas. 1988. *The Dream of the Earth*. San Francisco, CA: Sierra Club Books.

Berry, Wendell. 1981. *The Gift of Good Land*. San Francisco, CA: North Point Press.

Berry, Wendell. 1990. *What are People For?* San Francisco, CA: North Point Press.

Berry, Wendell. 1993. *Sex, Economy, Freedom and Community*. New York: Pantheon Books.

Berry, Wendell. 2008. "The Way of Ignorance," in Bill Vitek and Wes Jackson, eds, *The Virtues of Ignorance: Complexity, Sustainability, and the Limits of Knowledge*. Lexington, KY: University of Kentucky Press, pp. 37–50.

Bollier, David. 2014. *Think like a Commoner: A Short Introduction to the Life of the Commons*. Gabriola Island, BC: New Society Publisher.

Bonnett, Michael. 2003. "Issues for Environmental Education," *Journal of Philosophy of Education* 37(4): 691–705.

Borrelli, Peter, ed. 1988. *Crossroads: Environmental Priorities for the Future*. Washington, DC: Island Press.

Boström, Magnus. 2012. "A Missing Pillar? Challenges in Theorizing and Practicing Social Sustainability," *Sustainability Science, Practice, and Policy* 8(1): 3–14.

Bowers, C. A. 1995. *Educating for an Ecologically Sustainable Culture*. Albany, NY: State University of New York Press.

Boyce, J., A. Klemer, P. Templet, and C. Willis. 1999. "Power Distribution, the Environment, and Public Health: A State-level Analysis," *Ecological Economics* 29: 127–40.

Brock, Lothar. 1992. "Security Through Defending the Environment: An Illusion?" in Elise Boulding, ed., *New Agendas for Peace Research: Conflict and Security Reexamined*. Boulder, CO: Lynne Rienner, pp. 79–102.

Brower, David. 1995. *Let the Mountains Talk, Let the Rivers Run: A Call to Those Who Would Save the Earth*. New York: Harper Collins.

Brown, Lester R. 1981. *Building a Sustainable Society*. New York: W. W. Norton.

Brown, Lester R. 2009. *Plan B 4.0: Mobilizing to Save Civilization.* New York: W. W. Norton.

Brown, Lester R. 2011. *World on the Edge: How to Prevent Environmental and Economic Collapse.* New York: Earth Policy Institute.

Bullard, Robert. 1990. *Dumping in Dixie: Race, Class and Environmental Quality. Boulder.* Boulder, CO: Westview Press.

Bullard, Robert, ed. 1994. *Unequal Protection: Environmental Justice and Communities of Color.* New York: Random House.

Burke, Edmund. 1961. *Reflections on the Revolution in France.* Garden City, NY: Doubleday.

Carson, Rachel. 1962. *Silent Spring.* Boston, MA: Houghton Mifflin.

Carson, Rachel. 1998. "The Real World Around Us," in Linda Lear, ed., *Lost Woods,* Boston, MA: Beacon Press, pp. 148–63.

Carson, Rachel. 1998. *The Sense of Wonder.* New York: Harper.

Chertow, Marian R. 2000. "The IPAT Equation and Its Variants: Changing Views of Technology and Environmental Impact," *Journal of Industrial Ecology* 4(4): 13–29.

Clapp, Jennifer. May 2002. "What the Pollution Havens Debate Overlooks," *Global Environmental Politics* 2(2): 11–19.

Colburn, Theo, Dianne Dumanoski, and John Peter Meyers. 1997. *Our Stolen Future.* New York: Plume.

Commoner, Barry. 1972. *The Closing Circle.* New York: Bantam.

Commoner, Barry. 1990 (1975). *Making Peace with the Planet.* New York: Pantheon Books.

Commoner, Barry. 1994. "How Poverty Breeds Overpopulation (and Not the Other Way Around)," in Carolyn Merchant, ed., *Ecology: Key Concepts in Critical Theory,* Atlantic Highlands, NJ: Humanities Press, pp. 88–95.

Confucius. 1979. *The Analects.* New York: Penguin.

Corcoran, Peter Blaze, Kim E. Walker, and Arjen Wals. 2004. "Case Studies, Make-your-case Studies, and Case Stories: A Critique of Case-study Methodology in Sustainability in Higher Education," *Environmental Education Research* 10: 7–21.

Cordonier Segger, Marie-Claire and C. G. Weeramantry, eds. 2005. *Sustainable Justice.* The Netherlands: Martinus Nijhoff.

Costanza, Robert et al. 1997. "Case Studies, Make-your-case Studies, and Case Stories: The Value of the World's Ecosystem Services and Natural Capital," *Nature* 387 (May): 253–60.

Daily, Gretchen, ed. 1997. *Nature's Services: Societal Dependence on Natural Ecosystems.* Washington, DC: Island Press.

Daily, Gretchen et al. 2000. "The Value of Nature and the Nature of Value," *Science* 289 (July): 395–6.

Dalby, Simon. 2009. *Security and Environmental Change.* Cambridge: Polity.

Daly, Herman E. 1973. "The Steady-state Economy: Toward a Political Economy of Biophysical Equilibrium and Moral Growth," in Herman Daly, ed., *Toward a Steady-State Economy.* San Francisco, CA: W. H. Freeman, pp. 149–74.

Daly, Herman E. 1996. *Beyond Growth: The Economics of Sustainable Development.* Boston, MA: Beacon Press.

Deaton, Angus. 2008. "Income, Health and Wellbeing around the World: Evidence from the Gallup World Poll," *Journal of Economic Perspectives* 22: 53–77.

Diamond, Jared. 2005. *Collapse: How Societies Choose to Fail or Succeed.* New York: Viking.

Dowie, Mark. 2009. *Conservation Refugees: The Hundred-Year Conflict between Global Conservation and Native Peoples.* Boston, MA: MIT Press.

Dryzek, John S. 1997. *The Politics of the Earth: Environmental Discourses.* Oxford: Oxford University Press.

The Economist. March 3, 2012. "Free Exchange: Decline and Small," p. 87.

The Economist. March 24, 2012. "Free Exchange: Nudge Nudge, Think Think," p. 78.

The Economist. May 19, 2012. "Schumpeter: Good Business; Nice Beaches," p. 76.

Ehrlich, Paul R. 1968. *The Population Bomb.* New York: Ballantine Books.

Ehrlich, Paul R. 1991. "Population Diversity and the Future of Ecosystems," *Science* 254 (1991): 175.

Ehrlich, Paul R. and Anne Ehrlich. 1981. *Extinction: The Causes and Consequences of the Disappearance of Species.* New York: Random House.

Ehrlich, Paul R., John P. Holdren, and Barry Commoner. May 1972. "A Bulletin Dialogue: Critique and Response," *Bulletin of the Atomic Scientists* 28(5): 16–27, 42–56.

Esty, Daniel C. and Andrew Winston. 2009. *Green to Gold: How Smart Companies Use Environmental Strategy to Innovate, Create Value, and Build Competitive Advantage.* Hoboken, NJ: Wiley.

Finger, Matthias. 1994. "From Knowledge to Action? Exploring the Relationships between Environmental Experiences, Learning, and Behavior," *Journal of Social Issues* 50(3): 141–60.

Ford, Liz. 2012. "Rio+20 Politicians Deliver 'New Definition of Hypocrisy' Claim NGOs," *The Guardian*, June 21; accessed at http://www.guardian.co.uk/global-development/2012/jun/21/rio20-politicians-hypocrisy-ngos?intcmp=239.

Fountain, Henry. 2012. "Too Many Small Fish are Caught, Report Says," *New York Times*, April 2, p. A9.

French, Hilary F. 1997. "Learning from the Ozone Experience," in Lester R. Brown et al., eds, *State of the World 1997*. New York: W. W. Norton, pp. 151–71.

Friedman, Thomas L. 2006. *The World Is Flat*. New York: Farrar, Straus and Giroux.

George, Susan. 2010. *Whose Crisis, Whose Future? Towards a Greener, Fairer, Richer World*. Cambridge: Polity.

Gore, Al. 1992. *Earth in the Balance*. Boston, MA: Houghton Mifflin.

Gregory, R., D. Ohlson and J. Arvai. 2006. "Deconstructing Adaptive Management: Criteria for Applications to Environmental Management," *Ecological Applications* 16: 2411–25.

Hardin, Garrett. 1968. "The Tragedy of the Commons," *Science* 162: 1243–8.

Hardin, Garrett. 1993. *Living within Limits: Ecology, Economics, and Population Taboos*. New York: Oxford University Press.

Harris, Gardiner. 2012. "Prescription Rule Will Cut Use of Antibiotics in Livestock, F.D.A. Says," *New York Times*, April 12, p. A17.

Hartford, Tim. 2011. *Adapt: Why Success Always Starts with Failure*. New York: Farrar, Straus and Giroux.

Hawken, Paul. 1993. *The Ecology of Commerce: A Declaration of Sustainability*. New York: HarperCollins.

Hawken, Paul, Amory Lovins, and L. Hunter Lovins. 1999. *Natural Capitalism*. New York: Little, Brown.

Heinberg, Richard. 2004. *Powerdown: Options and Actions for a Post-Carbon World*. Gabriola Island, BC: New Society Publishers.

Heinberg, Richard. 2011. *The End of Growth: Adapting to Our New Economic Reality*. Gabriola Island, BC, Canada: New Society Publishers.

Heinberg, Richard and Daniel Lerch, eds. 2010. *The Post Carbon Reader*. Healdsburg, CA: Watershed Media.

Hern, Warren. 1993. "Has the Human Species Become a Cancer on the Planet? A Theoretical View of Population Growth as a Sign of Pathology," *Current World Leaders* 36(6) (December): 1089–1124.

Hines, Jody M., Harold R. Hungerford, and Audrey N. Tomera. 1986–87. "Analysis and Synthesis of Research on Responsible Environmental Behavior: A Meta-Analysis," *Journal of Environmental Education* 18 (Winter): 1–8.

Homer-Dixon, Thomas F. 1994. "Environmental Scarcities and Violent Conflict: Evidence from Cases," *International Security* 19(1): 5–40.

Homer-Dixon, Thomas F. 2001. *Environment, Scarcity, and Violence*. Princeton, NJ: Princeton University Press.

Homer-Dixon, Thomas F. 2006. *The Upside of Down: Catastrophe, Creativity, and the Renewal of Civilization*. Toronto: Knopf.

Hungerford, Harold R. and Trudi L. Volk. 1990. "Changing Learner Behavior through Environmental Education," *Journal of Environmental Education* 21(3) (Spring): 8–21.

International Union for the Conservation of Nature and Natural Resources. 1980. *World Conservation Strategy*. Gland, Switzerland: IUCN.

IUCN, United Nations Environment Programme, World Wide Fund for Nature. 1991. *Caring for the Earth: A Strategy for Sustainable Living*. Gland, Switzerland. https://portals.iucn.org/library/efiles/edocs/CFE-003.pdf.

Jackson, Tim. 2009. *Prosperity without Growth: Economics for a Finite Planet*. London: Earthscan.

Jacobsen, Kathryn H. 2008. *Introduction to Global Heath*. Boston, MA: Jones and Bartlett.

Jones, Holly and Oswald Schmitz. 2009. "Rapid Recovery of Damaged Ecosystems," *PLoS One* 4(5) (May): e5653.

Kahn, Herman, William Brown, and Leon Martel. 1976. *The Next 200 Years: A Scenario for America and the World*. New York: William Morrow and Company.

Kellert, Stephen. 2006. *Building for Life: Designing and Understanding the Human–Nature Connection*. Washington, DC: Island Press.

Kellert, Stephen and Edward O. Wilson, eds. 1993. *Biophilia Hypothesis*. Washington, DC: Island Press.

Kempton, Willett, James Boster, and Jennifer Hartley. 1995. *Environmental Values in American Culture*. Cambridge, MA: MIT Press.

Kibert, C., J. Sendzimir, and B. Guy. 2000. "Construction Ecology and Metabolism: Natural System Analogues for a Sustainable Built Environment," *Construction Management & Economics* 18: 903–16.

King Jr, Martin Luther. 1991. "Letter from Birmingham City Jail," in James Melvin Washington, ed., *A Testament of Hope: The Essential Writings and Speeches of Martin Luther King, Jr*. San Francisco, CA: Harper, pp. 289–302.

Knaup, Amy E. May 2005. "Survival and Longevity in the Business Employment Dynamics Data," *Monthly Labor Review* 12(5): 50–6.

Koger, Susan M. and Deborah Du Nann Winter. 2010. *The Psychology of Environmental Problems*, 3rd edn. New York: Taylor and Francis.

Kumar, Pushpamm, ed. 2010. *The Economics of Ecosystems and Biodiversity: Ecological and Economic Foundations*. London: Earthscan.

Lao-tzu. 1988. *Tao Te Ching*, trans. Stephen Mitchell. New York: HarperCollins.

Lappé, Frances Moore. 2010. *Getting a Grip 2*. Cambridge, MA: Small Planet Media.

Layard, R., G. Mayraz, and S. Nickell. 2008. "The Marginal Utility of Income," *Journal of Public Economics* 92: 1846–57.

Lee, Kai N. 1993. *Compass and Gyroscope: Integrating Science and Politics for the Environment*. Washington, DC: Island Press.

Leopold, Aldo. 1966. *A Sand County Almanac, with Essays on Conservation from Round River*. New York: Ballantine Books.

Lewin, R. 1986. "In Ecology, Change Brings Stability," *Science* 234: 1071–3.

Linn, Susan. 2004. *Consuming Kids: The Hostile Takeover of Childhood*. New York: New Press.

Lohmann, Larry. 1990. "Whose Common Future," *The Ecologist* 20(3) (May/June): 82–4.

Louv, Richard. 2005. *Last Child in the Woods: Saving our Children from Nature-Deficit Disorder*. Chapel Hill, NC: Algonquin Books.

Lovins, Amory and Rocky Mountain Institute. 2011. *Reinventing Fire: Bold Business Solutions for the New Energy Era*. White River Junction, VT: Chelsea Green Publishing.

Lovins, L. Hunter and Boyd Cohen. 2011. *Climate Capitalism: Capitalism in the Age of Climate Change*. New York: Hill and Wang.

MacIntyre, Alasdair. 1981. *After Virtue: A Study in Moral Theory*. Notre Dame, IN: University of Notre Dame Press.

Mackenzie-Mohr, Doug and William Smith. 1999. *Fostering Sustainable Behavior*. Gabriola Island, BC: New Society Publishers.

Marsh, George Perkins. 1869. *Man and Nature: Physical Geography as Modified by Human Action*. New York: C. Scribner's Sons.

Martin, Peter. 1996. "A WWF View of Education and the Role of NGOs," in John Huckle and Stephen Sterling, eds, *Education for Sustainability*. London: Earthscan, pp. 40–51.

Marx, Karl and Friedrich Engels. 1978. "Manifesto of the Communist Party," in Robert C. Tucker, ed., *The Marx-Engels Reader*, 2nd edn. New York: W. W. Norton, pp. 469–500.

McCloskey, Deirdre. 2006. *The Bourgeois Virtues: Ethics for an Age of Commerce*. Chicago, IL: University of Chicago Press.

McDonough, William and Michael Braungart. 2002. *Cradle to Cradle*. New York: North Point Press.

Meadows, Donella. 1982. "Whole Earth Models and Systems," *The Coevolution Quarterly* 34 (Summer): 98–108.

Meadows, Donella H. 2008. *Thinking in Systems*. White River Junction, VT: Chelsea Green Publishing.

Meadows, Donella, Dennis Meadows, Jorgen Randers, and William Behrens. 1972. *The Limits to Growth*. New York: Universe Books.

Meadows, Donella, Jorgen Randers, and Dennis Meadows. 2004. *Limits to Growth: The 30-Year Update*. White River Junction, VT: Chelsea Green Publishing.

Milbrath, Lester W. 1989. *Envisioning a Sustainable Society*. Albany, NY: State University of New York Press.

Mill, John Stuart. 1965. *Principles of Political Economy*, Volumes II–III in V. Bladen and J. Robson, eds, *The Collected Works of John Stuart Mill*. Toronto: University of Toronto Press.

Millennium Ecosystem Assessment. 2005. *Ecosystems and Human Well-Being, Synthesis*. Washington, DC: Island Press.

Mittelstaedt, John D., Clifford J. Shultz II, William E. Kilbourne, and Mark Peterson. 2104. "Sustainability as Megatrend: Two Schools of Macromarketing Thought," *Journal of Macromarketing* 34(3): 253–64.

Mollison, Bill. 1990. *Permaculture: A Practical Guide for a Sustainable Future*. Washington, DC: Island Press.

Mora, Camilo, Derek Tittensor, Sina Adi, Alastair Simpson, and Boris Worm. 2011. "How Many Species Are There on Earth and in the Ocean?" *PLoS Biology* 9(8) (August): e1001127.

Myers, Norman. 1993. *Ultimate Security: The Environmental Basis of Political Stability*. Washington, DC: Island Press.

New Jerusalem Bible. 1985. New York: Doubleday.

Nickerson, Mike. 2006. *Life, Money and Illusion: Living on Earth as if We Want to Stay*. Lanark, Ontario: Seven Generations Publishing.

Norgaard, Richard. 1992. "Coevolution of Economy, Society and Environment," in Paul Ekins and Manfred Max-Neef, eds, *Real-life Economics*. London: Routledge, pp. 76–88.

Northwest Earth Institute. 2010. *Choices for Sustainable Living*. Portland: NEI.

Ophuls, William and A. Stephen Boyan, Jr. 1992. *Ecology and the Politics of Scarcity Revisited: The Unraveling of the American Dream*. New York: W. H. Freeman.

Orr, David W. 1994. *Earth in Mind: On Education, Environment, and the Human Prospect*. Washington, DC: Island Press.

Orr, David. W. 2002. *The Nature of Design: Ecology, Culture, and Human Intention*. Oxford: Oxford University Press.

Ostrom, Elinor. 1990. *Governing the Commons: The Evolution of Institutions for Collective Action*. Cambridge: Cambridge University Press.

Owen, David. 2011. *The Conundrum: How Scientific Innovation, Increased Efficiency, and Good Intentions Can Make our Energy and Climate Problems Worse.* New York: Riverhead Books.

Parkin, Sara. 2010. *The Positive Deviant: Sustainability Leadership in a Perverse World.* London: Earthscan.

Partridge, Ernest. 1981. "Why Care about the Future," in Ernest Partridge, ed., *Responsibilities to Future Generations.* Buffalo, NY: Prometheus Books, pp. 203–20.

Peterson, Thomas C., Peter A. Stott, and Stephanie Herring, eds. 2012. "Explaining Extreme Events of 2011 from a Climate Perspective," National Oceanographic and Atmospheric Administration; accessed at: http://www1.ncdc.noaa.gov/pub/data/cmb/bams-sotc/2011-peterson-et-al.pdf.

Pimm, Stuart. 1991. *The Balance of Nature?* Chicago, IL: University of Chicago Press.

Pinchot, Gifford. 1910. *The Fight for Conservation.* New York: Doubleday, Page and Company.

Plato. 1989. *The Collected Dialogues.* Princeton, NJ: Princeton University Press.

Pollan, Michael. 2009. "Farmer in Chief," *New York Times Magazine*, October 12; accessed at: http://www.nytimes.com/2008/10/12/magazine/12policy-t.html?pagewanted=all.

Ponting, Clive. 1991. *A Green History of the World: The Environment and Collapse of Great Civilizations.* New York: St Martin's Press.

Putnam, Robert. June 2007. "E Pluribus Unum: Diversity and Community in the Twenty-first Century. The 2006 Johan Skytte Prize Lecture," *Scandinavian Political Studies* 30: 137–74.

Rawls, John. 1999. *A Theory of Justice*, revsd edn. Cambridge, MA: Harvard University Press.

Rawls, John. 2005. *Political Liberalism.* New York: Columbia University Press.

Rieckmann, Marco. 2012. "Future-oriented Higher Education: Which Key Competencies Should Be Fostered through University Teaching and Learning?" *Futures* 44(2) (March): 127–35.

Rockström, Johan et al. 2009. "A Safe Operating Space for Humanity," *Nature* 461 (September): 472–5.

Sachs, Aron. 1996. "Upholding Human Rights and Environmental Justice," in Lester R. Brown et al., eds, *State of the World 1996.* New York: W. W. Norton, pp. 133–51.

Sale, Kirkpatrick. 1985. *Dwellers in the Land: The Bioregional Vision.* New York: Random House.

Sandel, Michael J. 2012. *What Money Can't Buy: The Moral Limits of Markets.* New York: Farrar, Straus, and Giroux.

Schmidheiny, Stephan. 1992. *Changing Course: A Global Business Perspective on Development and the Environment*. Cambridge, MA: MIT Press.

Schor, Juliet. 2011. *True Wealth*. New York: Penguin.

Selby, David. July 2006. "The Firm and Shaky Ground of Education for Sustainable Development," *Journal of Geography in Higher Education* 30(2): 351–65.

Seufert, Verena, Navin Ramankutty, and Jonathan A. Foley. 2012. "Comparing the Yields of Organic and Conventional Agriculture," *Nature* 485 (May 10): 229–32.

Shermer, Michael. 2008. *The Mind of the Market*. New York: Holt.

Shiva, Vandana. 2005. *Earth Democracy: Justice, Sustainability, and Peace*. Cambridge, MA: South End Press.

Shrader-Frechette, Kristin. 2002. *Environmental Justice: Creating Equality, Reclaiming Democracy*. Oxford: Oxford University Press.

Sia, Archibald P., Harold R. Hungerford, and Audrey N. Tomera. 1985/1986. "Selected Predictors of Responsible Environmental Behavior: An Analysis," *Journal of Environmental Education* 17 (Winter): 31–40.

Simon, Julian. 1981. *The Ultimate Resource*. Princeton, NJ: Princeton University Press.

Singh, R. P. 2006. "Causes and Consequences of the Greenhouse Gases," *Bulletin of Pure and Applied Sciences – Geology* 25 (January): 1–2.

Smith, Gregory A. 1992. *Education and the Environment: Learning to Live with Limits*. Albany, NY: State University of New York Press.

Smith, Kirk, Carlos Corvalan, and Tord Kjellstroem. 1999. "How Much Global Ill Health is Attributable to Environmental Factors?," *Epidemiology* 10 (September): 573–84.

Smith, Laurence C. 2011. *The World in 2050*. New York: Penguin.

Solt, Frederick. 2008. "Economic Inequality and Democratic Political Engagement," *American Journal of Political Science* 52 (January): 48–60.

Soulé, Michael. 1980. "Thresholds for Survival: Maintaining Fitness and Evolutionary Potential," in Michael Soulé and Bruce Wilcox, eds, *Conservation Biology: An Evolutionary-Ecological Perspective*. Sunderland, MA: Sinauer Associates, pp. 111–24.

Starke, Linda and Lisa Mastny, eds. 2010. *2010 State of the World: Transforming Cultures*. Washington, DC: Worldwatch Institute.

Steffen, Will et al. 2015. "Planetary Boundaries: Guiding Human Development on a Changing Planet," *Science* 13: 347(6223). DOI: 10.1126/science.1259855.

Sterberg, Esther. 2009. *Healing Spaces: The Science of Place and Well-Being*. Cambridge, MA: Belknap Press.

Szabo, Liz. 2011. "Pregnant Women Rife with Chemicals," *USA Today*, January 14, p. 3A.

Tainter, Joseph A. 1988. *The Collapse of Complex Societies*. Cambridge, UK: Cambridge University Press.

Taylor, Bron. 2008. *Encyclopedia of Religion and Nature*. London: Continuum.

Thaler, Richard H. and Cass R. Sunstein. 2008. *Nudge: Improving Decisions about Health, Wealth, and Happiness*. New Haven, CT: Yale University Press.

Thiele, Leslie Paul. 1999. *Environmentalism for a New Millennium: The Challenge of Coevolution*. New York: Oxford University Press.

Thiele, Leslie Paul. 2011. *Indra's Net and the Midas Touch: Living Sustainably in a Connected World*. Cambridge, MA: MIT Press.

Thornton, Joe. 2000. *Pandora's Poison*. Cambridge, MA: MIT Press.

Torras, Mariano and James Boyce. 1998. "Income, Inequality, and Pollution: A Reassessment of the Environmental Kuznets Curve," *Ecological Economics* 25: 147–60.

Tremmel, Joerg Chet, ed. 2006. *Handbook of Intergenerational Justice*. Cheltenham, UK: Edward Elgar.

Union of Concerned Scientists. 2012. "Smarter Ways to Keep the Planet Cool," *Catalyst* (Spring): 7–8.

United Church of Christ Commission for Racial Justice. 1987. *Toxic Wastes and Race in the United States*. New York: United Church of Christ Commission for Racial Justice.

Wals, Arjen. 2009. *Review of Contexts and Structures for Education for Sustainable Development: Learning for a Sustainable World*. Paris: UNESCO.

Wals, Arjen and Bob Jickling. 2002. "Sustainability in Higher Education: From Doublethink and Newspeak to Critical Thinking and Meaningful Learning," *Higher Education Policy* 15: 121–31.

Walters, C. and C. S. Holling. 1990. "Large-scale Management Experiments and Learning by Doing," *Ecology* 71: 2060–8.

Weiss, Edith Brown. 1989. *In Fairness to Future Generations*. Tokyo: The United Nations University.

Wenz, Peter. 2007. "Does Environmentalism Promote Injustice for the Poor?" in Ronald Sandler and Phaedra Pezzulo, eds, *Environmental Justice and Environmentalism: The Social Justice Challenge to the Environmental Movement*. Cambridge, MA: MIT Press, pp. 57–84.

White, Jr, Lynn. 1967. "The Historical Roots of Our Ecologic Crisis," *Science* 155: 1203–7.

Wilkinson, Richard and Kate Pickett. 2009. *The Spirit Level: Why Greater Equality Makes Societies Stronger*. New York: Bloomsbury Press.

Wilson, Edward O. 1984. *Biophilia*. Cambridge, MA: Harvard University Press.

Wilson, Edward O. 2002. *The Future of Life*. New York: Vintage Books.

World Commission on Environment and Development. 1987. *Our Common Future*. Oxford: Oxford University Press.

World Commission on the Ethics of Scientific Knowledge and Technology. 2005. *The Precautionary Principle*. Paris: UNESCO.

Worster, Donald. 1995. "The Shaky Ground of Sustainability," in George Sessions, ed., *Deep Ecology for the 21st Century*. Boston, MA: Shambhala, pp. 417–27.

Yorque, Ralf, et al. 2002. "Toward an Integrative Synthesis," in Lance H. Gunderson and C. S. Holling, eds, *Panarchy: Understanding Transformations in Human and Natural Systems*. Washington, DC: Island Press, pp. 419–38.

Index